Soup Alive!

Soup Alive!

By Eleanor S. Rosenast

Woodbridge Press
Santa Barbara, California 93160

Published and distributed by

Woodbridge Press Publishing Company
Post Office Box 6189
Santa Barbara, California 93160

Distributed simultaneously in the United States and Canada.

Printed in the United States of America

Library of Congress Cataloging-in-Publication data:

Rosenast, Eleanor S.

Soup alive! / by Eleanor S. Rosenast.

    p.   cm.

Includes index.

ISBN 0-88007-198-2 : $9.95

    1. Vegetarian cookery. 2. Soups. 3. Raw foods.
      4. Enzymes.   I. Title

TX757.R728 1993

641.8'13—dc20   93-6656
                                     CIP

Cover paintings: Janice Blair
Text illustrations: Corel Corporation, CD-ROM Clipart, CorelDraw 3.0;
   and Totem Graphics.

Dedication

To Marion Udry, who showed me there is a way;
and my husband, Hans, who helps me to stay with it.

Acknowledgments

I did not write this book alone. I thank my family and friends who served as tasters. Their comments and ideas for soup helped create the book.

I am especially grateful to Susan Hinerfeld and Annice Jacoby for editing the manuscript. Without their encouragement and valuable suggestions, *Soup Alive!* would still be in my desk drawer!

Contents

Introduction   7
  How to Save the Enzymes   9
  Observe Food Safety   10
  Special Ingredient Concerns   10
  A Time Saver   11
  Food Processor or Blender   12
  Raw Vegetables   12
  Soup Base   13
  Thickeners   15
  Seasonings   15
  Garnishes   16

List of Recipes   17

The Recipes 23 - 172

Introduction

Diet awareness is the hot topic of the decade. Changes in eating and cooking styles are taking place. Health conscious consumers are looking for nutritious recipes. High fiber vegetables prepared to their best advantage are linked to a healthy heart and to success in cancer prevention by some medical authorities. The foods we choose now need to enhance the life we lead.

I do not take up my task of reporting the marvel of a new soup tradition as a famous chef, learned nutritionist or medical researcher. I have gathered my information because a serious threat to my life made me evaluate what my diet was actually giving me. I am writing as a consumer, concerned with quality of life, an educator, a cook, a wife, a parent and, of course, an eater.

During my search for health after a bout with cancer, the idea for this salubrious soup cookbook, *Soup Alive!*, evolved. With my school teacher thoroughness, I went exploring for possible lesson plans in nutrition. I heeded the advice of the National Cancer Institute, the American Heart Association and the National Institutes of Health to follow a low-fat, high-fiber diet. The N.C.I.'s suggestions for eating cruciferous vegetables like Brussels sprouts, broccoli, cabbage and cauliflower in order to decrease the risk of breast and colon cancer seemed to make sense too.

We are surrounded by reports on various methods for achieving optimum health. In my campaign to stay healthy, I have tried a low-fat diet, megavitamins, large doses of carrot juice, wheat grass juice, a vegetarian diet, a raw food diet, etc. Sometimes, in a whimsical mood, I categorize them as Diet Most Likely to Succeed, Easiest to Stick With, Most Enjoyable, Most Distasteful, Most Difficult to Follow—not to overlook Most Expensive.

Through all these plans for health there runs a similar theme: *Give Your Body Optimum Nutrition*. The authorities agree with Popeye and your mother, eat your spinach and fight to the finish. An apple a day keeps the doctor away.

This movement to include more fruits and vegetables in your diet emphasizes eating them in the most natural state possible. Accumulated scientific evidence suggests that vegetables eaten raw or lightly cooked supply more dietary vitamins and enzymes.

Enzymes are found in the food you eat. They convert the food into chemical structures that can pass through the cell membranes. Enzymes aid in converting food into muscle, flesh, bone, nerves and glands. They break down the millions of cells in plants and animals. All uncooked food contains an abundance of enzymes which correspond to the nutritional highlights of the food. For example, food high in fat contains concentrations of the enzyme lipase which aids in the digestion of fat. Fruits and vegetables contain large quantities of the enzyme cellulase which is needed to break down plant fibers. All raw food has the correct and balanced amounts of food enzymes for human digestion.

Life could not exist without enzymes, but scientists agree that enzymes are highly fragile. They break down from excessive light and pressure—but especially from heat.

Does cooking your food leave you with food without enzymes?

Could a diet low in enzymes be the culprit in many human ailments? Is this the underlying cause of some health problems? Why take a chance? Include more raw food in your diet. Develop a preventive medical program in your kitchen. Use your soup spoon and the *Soup Alive!* cookbook to enhance your health.

In my attempt to eat a raw vegetarian diet and to keep peace in my family, the idea for this book was conceived. My Swiss husband's quest for health is not as arduous as mine. His taste buds were educated in the land of "rosti" (fried potatoes), "bratwurst" (sausage) and overcooked vegetables. He has little taste for vegetables that are still in a state of nature. To change his pouting face at the dinner table, I searched for a way to make the raw vegetable diet tasty. In spring and summer eating raw food is inviting. Soups and salads made from raw vegetables and fruits are refreshing. But in colder weather hot foods are more satisfying. Serving the living soups hot helped make the raw food diet more savory and soon my husband was asking for seconds.

How to Save the Enzymes!

The approach to making hot soups in *Soup Alive!* is not to cook the vegetables! Prepare the fresh vegetables—peeling, shredding, cutting or chopping—as called for in the individual recipes. The prepared vegetables, along with seasonings and thickeners, are added to the appropriate soup base. Always warm the soup slowly on low heat, with the lid off. This will give you more control. Stir frequently. Do not allow the soup to reach a simmer.

Another way to warm the soup without long cooking is to use a double boiler or follow the French method and create a "Bain Marie." Fill a heat resistant glass container with the soup and place the container in hot water until the soup is warmed; or, the soup-filled "Bain Marie" can be placed in the oven and heated there. These methods prevent direct contact with the heat source and protect the enzymes in the soup. You achieve the comfort of a hot bowl of soup with little loss of nutrition. The more raw the food, the more intact the vitamins and enzymes and the healthier you will be.

Observe Food Safety

Please Note: **Because these soups are not boiled, special attention should be paid to any leftovers. Do not leave the remaining soup uncovered on the counter for long periods, or unrefrigerated. To insure pristine taste, nutrition and safety, refrigerate the remaining soup as soon as possible.**

Special Ingredient Concerns

Vegetarian cooking is in a state of flux. Current opinions vary as to which ingredients are most healthful. However, mainstream ideas are in agreement on the need to have more fruits and vegetables in the diet and on the benefits of eating them in their most natural state.

Vegetarians even debate the efficacy of garlic and onions. If this is one of your areas of concern, try asafoetida. Asafoetida is a spice used in Indian and Middle Eastern cooking. One-eighth teaspoon of asafoetida powder is approximately equivalent in taste to 1/4 cup chopped onion or one clove of garlic. Asafoetida is available in the spice section of most health food stores, or you can find it in a Middle Eastern market. If you have trouble locating this spice, you can order it from: Frontier Herbs, P.O. Box 118, Norway, IA 52318, (800)365-4372.

The *Soup Alive!* recipes offer the category "Milk product of choice." If you are concerned about your intake of animal fat, you can use evaporated skim milk or soy milk without radically changing the taste of the soup. Evaporated skim milk gives a creamy consistency without too much animal fat. If it is a once-in-a-while company dinner and you want the soup to be at its optimum, you might use cream. However, soy milk works very well and your guests may not suspect.

"Oil of choice" is another category offered in the recipes. If you are concerned about cholesterol or saturated fats, avoid the butter and use olive oil, canola oil or margarine.

A Time Saver

You will not only enjoy the recipes in *Soup Alive!* but you will find this method of preparing soup a great time saver. The boiling and endless simmering associated with making soup is eliminated. The recipes are especially convenient for solo cuisine. An on-the-go health conscious single who finds cooking a nuisance can with minimal effort fix a nutritious meal. These gourmet soups are fast and easy to prepare.

In addition to accommodating the lifestyle of singles, the recipes are also convenient for parents with an infant to feed—the texture of the soup can be controlled to make it suitable for a baby. While fixing soup for dinner, a parent can make baby food for a month. The leftover soup can be frozen in ice cube trays, defrosted and heated when needed. The recipes in *Soup Alive!* are a healthy, efficient and economical way to feed a baby as well as the parents.

All you need is:

Food Processor or Blender
Raw Vegetables
Soup Base
Thickeners
Seasonings
Garnishes
and the *Soup Alive!* cookbook

Food Processor or Blender

The food processor or blender is used to liquefy the vegetables. If you use a blender, start with the prepared vegetables and a small amount of soup base. The liquid should just barely cover the vegetables.

Start with the blender on high speed. This gets the solids across the blades before they make the puree too thick. High speed supplies the motor with more power. If you hit an air pocket, start again with the blender on slow. If this does not work, add additional liquid and start again. Add the least amount of liquid you can to get the blender going. If the mixture is too chunky and you do not want that consistency, use more liquid. Keep increasing the liquid a little at a time until the vegetables are pureed.

If you are using a food processor, use the chopping blade—also called the "S" blade. In a food processor, you can do the initial chopping of the vegetables without liquid. When the vegetables are chopped as fine as possible, start to add the liquid a little at a time with the food processor working until the puree has reached the consistency you desire.

Raw Vegetables

The raw vegetables used in the *Soup Alive!* recipes range from Avocado to Zucchini. Purchase vegetables of the best quality and as fresh as possible. Bouquets of crisp vegetables will ensure good taste. Buy vegetables that are in season and locally grown, if possible, rather than those that are shipped from great distances.

Prepare the vegetables by peeling and paring where appropriate. Then cut them to size to fit into your blender or food processor. Softer vegetables such as spinach, asparagus, tomatoes, summer squash, mushrooms, etc., make the task of pureeing easier and give a smoother texture. However, even raw carrots and beets puree to a soup-like texture.

Soup Base

Each recipe specifies a particular soup base. The base is mixed with the vegetable puree to further liquefy and create the texture of a soup. Some of the suggestions for soup base are simple and do not require much preparation, e.g., milk, soy milk, evaporated milk, cream, tomato juice, apple juice or a combination of these.

To some extent the size of the vegetables will depend on your personal taste and who is coming to dinner. Cloves of garlic vary in size; if you prefer a garlicky taste, use a thicker clove of garlic. If you prefer a thicker soup, your one bunch of spinach will be on the heavier side. You can always thin your soup with additional bouillon, soup base or water. If you have unexpected guests, follow my mother's advice: "More guests, more water."

For a more savory soup, prepare your own rich-tasting stocks in advance. Keep them on hand in the refrigerator or freezer. A basic stock can be prepared by saving a week's supply of vegetable parts ordinarily thrown away, e.g., peelings, tops, unused stalks, roots, shoots, etc. Also, save vegetable cooking water left over when you steam vegetables for other purposes. Wash the discards and add two quarts of vegetable cooking water. The ratio should be 2 cups of discarded vegetables to 2 quarts of water. Simmer for 1/2 hour and strain. Do not use too many strong vegetables such as onions, cabbage or turnips, as this will dominate the flavor of the stock. Commercially prepared powdered soup mix can be combined with the basic stock to further enrich the flavor. This basic stock can be used in making all of the raw vegetable soups.

Dried mushroom stock added to your fresh mushroom soup creates a dish for a gourmand. To make the stock take 1/2 cup dried black mushrooms and add them to 1 cup of boiling water. Remove the pan from the heat and let stand for 1 hour. Chop the mushrooms finely.

Tomato-flavored stock can be made following the same procedure. Use 1/2 cup of dried tomatoes. This stock is especially delicious when used in string bean soup, spinach soup and zucchini soups.

Onion-flavored stock made with the same procedure, substituting 1/2 cup of dried onion, will help to create a savory raw soup when used with any vegetable.

Garlic-flavored stock contributes a savory quality to soups made from tomatoes, spinach, zucchini, green beans and mixed vegetables. Press all of the cloves of an entire head of garlic in a garlic press, or mince the garlic fine. Sauté the pressed or minced garlic in 2 Tbsp oil of choice (olive, canola, margarine, butter). Add 5 cups of boiling water, salt and pepper to taste. Simmer for 30 minutes and allow to cool.

A vegetarian white stock can be made by adding 1 carrot, 2 leeks, 1 onion, 3 stalks of celery and 3 sprigs of parsley to 1 quart of water and simmer for 1/2 hour. For increased flavor stick the onion with a clove, or add a bay leaf while simmering. Strain and reserve the liquid and use in the raw soups.

A mock cream-of-chicken soup base can be made by making a roux with 1/4 cup butter or margarine and 1/2 cup rice flour. Cook the roux until it turns golden. Add 2 quarts vegetarian chicken flavored broth (can be purchased in a health food store), 1 onion, 1 stalk celery, 2 leeks, 2 sprigs of parsley, salt and pepper to taste and simmer for 30 minutes. Strain and use this soup base rather than dairy products in the cream soups.

Another effective way to create a cream soup base is to cook 2 cups of vegetarian chicken flavored boullion (or, if you choose, regular chicken boullion) with 2 tablespoons of pearl tapioca for 1/2 hour. This soup base is especially delicious in a cream-of-tomato soup or a cream-of-pea soup.

Thickeners

Each soup contains either oatmeal, farina, or a boiled potato. These ingredients thicken the pureed vegetable liquid, create a creamy texture and help to give the consistency of a cooked soup.

The recipes indicate which thickening ingredient to use and the appropriate amount. The thickening ingredient of choice is added to the vegetable puree. The best time to combine the puree and the thickener is when you are in the process of adding liquid and continuing to blend. This assures that the thickener will be well-blended with the soup.

Seasoning

Seasoning is the key to a savory soup in the *Soup Alive!* recipes. It is a challenge to get the zest of the seasoning to release in a raw soup. Meat-free liquid boullion made of vegetable protein derivatives is used as a flavor amplifier. Liquid boullion is more efficient than boullion cubes. It blends faster and more thoroughly without leaving any undissolved particles.

A variety of powdered soup mixes are used in the recipes. Onion soup mix, chicken flavored soup mix and vegetable soup mix add aroma, body, character and flavor to the raw soup. The powdered soup mix is added to the puree and mixed in the food processor or blender.

Dried and fresh herbs are used liberally. Basil, bay leaves, cayenne pepper, oregano, nutmeg, paprika and thyme are added to the appropriate soups. If the recipe calls for sautéed garlic and onions, the fresh or dried herbs are added to the sauté. If sautéing is not called for, add the herbs during the pureeing operation. For additional texture and aroma, fresh herbs can be chopped and sprinkled on the soup just before serving.

Tamari, a soy sauce, takes the place of salt in some recipes. It is made from soybeans and water. Tamari adds a piquant flavor to the soup. Some of the recipes call for Spike. Spike is a combination of blended herbs, dried vegetables and sea salt. Miso, used in some recipes, is made from fermented soy beans. It adds a rich flavor to the soup and is purported to be a source of beneficial bacteria that can aid digestion. These ingredients are available in most health food stores and super markets.

Always test the soup and adjust the seasoning to your taste. Do not depend entirely on the measurements in the recipes. They are only a guide. Add additional seasoning a little at a time, so as not to overwhelm the soup.

Garnishes

Garnishes are your *pièce de résistance.* Members of your household who are less enthusiastic about the new diet may be tempted by a tasty decorative garnish.

Dress up soups with varied accompaniments. Croutons, noodles, beans, rice, grated cheese, pesto, grated hard-cooked eggs, popcorn, cheese puffs, toasted garlic bread, shredded almonds and avocado slices create attractive soups and are also nutritious additions.

The goal of this book is to help you to eat more healthfully and to relish it. Ingredients for the recipes are available in supermarkets and health food stores. The average yield of the recipes is 4-5 servings. Singles doing solo cooking will find it easy to cut down the amounts.

The soups made from these recipes emphasize freshness, flavor and nutritional value. If you are already committed to the benefits of raw food, you will find the new combination of ingredients and simple-to-prepare recipes in the *Soup Alive!* cookbook an addition to your diet. Raw vegetable + soup base + thickener + seasoning + garnish = a nutritious delicious bowl of soup.

À Votre Santé!

Soup Alive!

Salubrious Soups from A to Z

(Recipes yield 4-5 servings)

Avocado Almond Soup   23
Avocado Chowder   24
Avocado Soup Cream Style   25
Avocado Soup Mexican Style   26
Asparagus Soup Almond Flavor   27
Asparagus Soup Simple   28
Asparagus Soup Cream Style   29

Basil Soup Simple   30
Basil Soup Italian Style   31
Basil Tomato Soup   32
Beet Sauerkraut Soup   33
Beet Soup Cream Style   34
Beet Soup Polish Style 35
Beet Soup Russian Style   36
Beet Soup Simple   37
Beet Soup Spicy Style 38
Beet Soup Zesty Style   39
Broccoli Soup Cream Style   40
Broccoli Soup Honey Mustard Flavor   41
Broccoli Soup Japanese Style   42
Broccoli Soup Scandinavian Style 43
Broccoli Soup Simple 44

Brussels Sprouts Soup Spicy Style   45
Brussels Sprouts Dahl Soup   50

Cabbage Tomato Soup   46
Cabbage Soup Swedish Style   47
Cabbage Soup Sweet-and-Sour   48
Cabbage Soup Zesty Style   49
Carrot Chowder   51
Carrot Coconut Soup   52
Carrot Dill Soup   53
Carrot Oatmeal Soup   54
Carrot Soup Curry Flavor   55
Carrot Peanut Butter Soup   56
Carrot Sunflower Seed Soup   57
Carrot Soup Orange Flavor   58
Cauliflower Chowder   59
Cauliflower Coconut Soup   60
Cauliflower Soup Cream Style   61
Cauliflower Soup Caribbean Style   62
Cauliflower Soup Thai Style   63
Cauliflower Soup Spicy Style   64
Celery Barley Soup   65
Celery Soup Simple   66
Celery Soup Cream Style   67
Celery Tomato Soup   68
Chard Soup Cream Style   69
Chard Soup Curry Flavor   70
Chard Soup Egyptian Style   71
Chard Soup Japanese Style   72
Chard Soup Jewish Style   73
Chard Soup Simple   74
Chard Tomato Soup   75
Corn Basil Soup   76
Corn Soup Confetti Style   77
Corn Soup Cream Style   78
Corn Soup Simple   79
Corn Soup Zesty Style   80

Corn Tomato Soup   81
Cucumber Dill Soup 82
Cucumber Soup Simple 83
Cucumber Beet Soup 84
Cucumber Tomato Soup 85

Fennel Soup Simple 86
Fennel Soup Italian Style 87
Fennel Soup Spicy Style 88

Green Bean Almond Soup   89
Green Bean Soup Anise Flavor   90
Green Bean Basil Soup   91
Green Bean Soup Simple   92
Green Bean Soup Cream Style 93
Green Bean Soup Tomato Flavor 94
Green Pea Asparagus Soup 95
Green Pea Soup Cream Style 96
Green Pea Soup Ginger Flavor 97
Green Pea Soup Oriental Style 98
Green Pea Sesame Seed Soup 99
Green Pea Soup Simple 100
Green Pea Soup Spicy Style 101
Green Pea Tomato Soup   102

Kale Basil Soup   103
Kale Soup Indian Style   104
Kale Soup Portuguese Style   105
Kohlrabi Almond Soup   106
Kohlrabi Soup Cream Style   107
Kohlrabi Soup Zesty Style   108
Kohlrabi Tomato Soup   109

Lettuce Celery Soup   110
Lettuce Green Pea Soup   111
Lettuce Soup Cream Style   112
Lettuce Soup Simple   113

Mushroom Kasha Soup   114

Mushroom Soup Chinese Style 115
Mushroom Orange Soup 116
Mushroom Tomato Soup   117
Mushroom Soup Russian Style 118
Mushroom Soup Simple   119
Mushroom Sesame Soup   120

Parsley Soup Simple   121
Parsley Soup Swiss Style   122
Parsley Carrot Soup   123
Parsnip Apricot Soup   124
Parsnip Soup Curry Flavor   125
Parsnip Soup Fennel Flavor   126
Parsnip Dill Soup   127
Parsnip Soup Cream Style   128
Parsnip Soup Simple   129
Parsnip Soup Spicy Style   130
Pumpkin Soup Seminole Style 131
Pumpkin Soup Zesty Style   132

Red Bell Pepper Fennel Soup   133
Red Bell Pepper Soup Cream Style   134
Red Bell Pepper Tomato Soup   135

Spinach Soup Cream Style   136
Spinach Soup Curry Flavor 137
Spinach Soup Italian Style   138
Spinach Soup Peanut Flavor   139
Spinach Soup Scandinavian Style   140
Spinach Soup Simple   141
Spinach Soup Pesto Flavor   142
Summer Squash Dill Soup   143
Summer Squash Soup Colombian Style   144
Summer Squash Soup Cream Style   145
Summer Squash Soup Simple   146
Summer Squash Soup Apple Flavor   147

Tomato Carrot Soup   148

Tomato Peanut Butter Soup   149
Tomato Pesto Soup 150
Tomato Soup Cream Style 151
Tomato Soup Onion Flavor   152
Tomato Soup Pineapple Flavor   153
Tomato Soup Saffron Flavor 154
Tomato Soup Simple   155

Vegetable Soup Basque Style 156
Vegetable Soup Japanese Style 157
Vegetable Soup Mexican Style   158
Vegetable Soup Swiss Style 159
Vegetable Soup Portuguese Style   160
Vegetable Soup Italian Style 161

Yam Soup Apple Flavor   162
Yam Almond Butter Soup   163
Yam Soup Indonesian Style   164
Yam Soup Simple   165
Yam Soup Cream Style   166

Zucchini Chowder   167
Zucchini Soup Simple   168
Zucchini Soup Italian Style   169
Zucchini Soup Mediterranean Style   170
Zucchini Soup Pistou Flavor   171
Zucchini Soup Spicy Style   172

Avocado Almond Soup

1 Tbsp. oil of choice (olive, canola, margarine, butter)
1 small onion
1 Tbsp. tomato paste
1/2 green pepper
2 stalks celery
1 medium tomato
1 large avocado
1/2 cup almond butter
3/4 Tbsp. lemon juice
4 cups vegetable bouillon
Salt and pepper to taste

Heat oil of choice in a soup pot. Chop onion fine and sauté until translucent. Add tomato paste. Cook one minute longer. Stir well.

Discard seeds and core from green pepper. Cut celery into 1" pieces. Chop ingredients in a food processor or blender.

Cut tomato and avocado into small chunks. Add them to the work bowl and process. Add almond butter, lemon juice and small amounts of vegetable bouillon to keep the blade moving. Add remaining bullion, pureeing until a soup-like consistency is achieved.

Remove soup to pot. Warm over low heat just below the simmer point. Stir often. Season to taste with salt and pepper. If you desire, soup can be eaten at room temperature.

Garnish: Thinly sliced, toasted almonds

Avocado Chowder

1 small onion
2 cloves garlic
1 Tbsp. oil of choice (olive, canola, margarine, butter)
1 tsp. marjoram
2 cups tomato juice (fresh or canned)
1 medium steamed potato
1 medium carrot
2 stalks celery
2 cups vegetable bouillon
1 tsp. basil
1 large avocado
Salt and pepper to taste

Heat oil of choice in a soup pot. Chop garlic and onion fine. Sauté until slightly brown. Add marjoram and basil. Stir well. Add a little of the tomato juice and cook a few minutes longer.

Cut carrot and celery into 1″ chunks. Chop in a food processor or blender. Peel avocado, add it and the steamed potato to the mixture and blend. Pour vegetable bouillon and remaining tomato juice, a little at a time, into the work bowl, continuing to run the machine until a soup-like consistency is achieved.

Transfer soup to pot and heat slowly. Do not allow the soup to come to a simmer. Stir frequently. Adjust the seasoning to your taste with salt and pepper.

Garnish: **Taco-brah**

Heat 2 tablespoons olive oil in a frying pan. Tear 6 tortillas into eighths and fry in the oil. Add 2 tablespoons of your favorite spaghetti sauce and 1/2 cup grated Monterey Jack cheese. If desired, soy cheese may be used.

Avocado Soup Cream Style

2 medium avocados
2 uncooked, medium potatoes
4 cups mock-chicken soup
   (see Introduction, or use chicken-flavored bouillon)
1/2 small onion
1 Tbsp. oil of choice (olive, canola, margarine, butter)
1 cup milk product of choice (cream, milk, evaporated,
   skim milk, soy milk—buttermilk is especially
   delicious in this recipe)
1 Tbsp. tamari

Quarter potatoes and cook them in two cups of mock-chicken soup or chicken-flavored bouillon, until they are very soft. Puree them in a food processor or blender and set aside until cool.

Chop onions fine and sauté until golden in oil of choice. Peel avocados and blend with cooled potato puree. Add remaining liquid and 1 cup of milk product of choice. Add 1 tablespoon tamari.

Transfer soup to a kettle and heat slowly. Do not allow the soup to come to a simmer. Stir frequently. Adjust the seasoning to your taste with additional tamari.

Garnish: Dust with paprika and fresh corn kernels cut off the cob.

Avocado Soup Mexican Style

3 small avocados
1 small onion
1 medium steamed potato
1 Tbsp. oil of choice (olive, canola, margarine, butter)
4 cups chicken-flavored bouillon
1 tsp. cumin*
1/2 tsp. chili powder*
Pinch of cayenne pepper*
Salt and pepper to taste

* Or use 1/4 tsp. mild salsa in place of all these spices.

Chop onion fine. Put in soup pot with oil of choice, cumin, chile powder and cayenne pepper. Cook until onions are soft. Stir frequently.

Peel avocados and chop in a food processor or blender. Add potato with small amounts of the chicken-flavored bouillon, continuing to work the machine until a soup-like consistency is formed. If using salsa in place of the spices, add the salsa during this process.

Heat the soup slowly, stirring frequently. Do not allow the soup to come to a simmer. Adjust the seasoning with salt and pepper.

Garnish: Corn chips

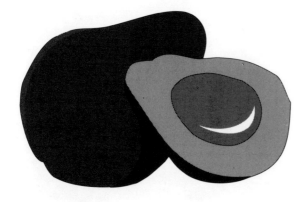

Asparagus Soup Almond Flavored

1 bunch asparagus (1 lb.)
2 celery stalks
1/2 cup ground almonds
4 cups vegetable bouillon
1/8 tsp. oregano
1/8 tsp. thyme
Salt and pepper to taste

Break tough ends off asparagus and save for stir frying or salads. Chop asparagus in a blender or food processor.

Cut celery into 2" pieces and add to the work bowl a few at a time. Use a little of the bouillon to help the blade go around.

Continue to puree the asparagus and celery, adding small amounts of the bouillon.

Add the ground almonds, oregano and thyme and work the machine until a soup-like consistency is achieved.

Remove the soup to a pot and heat slowly. Do not allow the soup to come to a simmer. Stir frequently. Season to taste with salt and pepper.

Garnish: Fresh corn kernels cut off the cob.

Asparagus Soup Simple

1 bunch asparagus
2 stalks celery
1 medium steamed potato
4 cups chicken-flavored bouillon
Gamasio to taste (sesame seeds and salt ground
    together, available at health food stores)

Break asparagus at the tough ends. Save the tough ends
for salads and stir frying. Cut asparagus into 1" lengths.

Chop the asparagus and celery in a food processor or
blender until fine. Add the potato. Continue to work the
machine, while adding small amounts of bouillon until a
soup-like consistency is formed.

Pour this mixture into a soup pot. Heat, but do not
allow the soup to come to a simmer. Stir frequently.
Season to taste with gamasio.

Garnish: **Sesame Toast**

1 Tbsp. sesame seeds for each slice of whole wheat
bread.

1/2 tsp. oil of choice (olive, canola, butter, margarine)
for each slice of whole wheat bread.

Toast sesame seeds in a dry frying pan until they are
browned. Stir frequently. Spread oil of choice on bread
and top with toasted sesame seeds. Bake in a 350° oven
for five minutes.

Asparagus Soup Cream Style

1/2 medium onion
2 stalks celery
1 tsp. dill
1 lb. asparagus
1 medium steamed potato
2 cups vegetable bouillon
2 cups milk product of choice (soy milk,
    evaporated skim milk, milk, cream)
Salt and pepper to taste

Chop onions and celery fine. Heat margarine or butter in a soup pot and add onions and celery. Sauté until golden brown. Add dill. Stir well.

Snap off tough ends of asparagus. Save for other recipes, e.g., salads, stir-frying. Chop asparagus spears, a few at a time, in the food processor or blender.

Add the steamed potato, continuing to work the machine. Add small amounts of the vegetable bouillon, pureeing until a soup-like consistency is achieved. Add milk product of choice and whirl the machine once again.

Transfer soup to pot. Warm slowly over low heat to serving temperature. Do not allow the soup to reach a simmer. Stir frequently. Season to taste with salt and pepper.

Garnish: Herbed croutons

Basil Soup Simple

3 cloves garlic
1 Tbsp. oil of choice (olive, canola, margarine, butter)
1/2 tsp. tarragon
1 bunch basil
1 medium steamed potato
2 cups chicken-flavored bouillon
2 cups tomato juice
Salt and pepper to taste

Mince garlic and sauté in oil of choice. Add tarragon and cook a little longer. Do not allow the garlic to brown. Stir often.

Remove the basil leaves and discard the stems. Should have approximately 1 cup leaves loosely stacked. Chop the leaves in a food processor or blender. Add the steamed potato and continue to work the machine while adding the liquid ingredients, a little at a time, until a soup-like consistency is formed.

Place the soup in a pot along with the garlic sauté. Heat on a low flame until warm. Do not allow the soup to come to a simmer. Stir often. Season to taste with salt and pepper.

Garnish: Garlic-flavored popcorn

Basil Soup Italian Style

8 green onions
3 cloves garlic
1 Tbsp. olive oil
1 tsp. sugar
1 tsp. salt
1 Tbsp. balsamic vinegar
2 small zucchini
1 small red pepper
3 medium tomatoes
3/4 cup basil leaves, firmly packed
1 small steamed potato
4 cups chicken-flavored bouillon

Remove most of the green end of the onions and save for salads or stir-frying. Slice the white end thin. Mince garlic. Add onion and garlic to olive oil and cook in a kettle until soft. Add balsamic vinegar, sugar and salt. Stir well.

Trim zucchini and cut into 1″ pieces. Chop in a food processor or blender. Discard seeds of red pepper. Add pepper to work bowl and process. Quarter tomatoes and potato and chop them with the rest of the ingredients. Add basil leaves, continuing to chop until the vegetables are assimilated. Use a little of the bouillon to keep the blade moving. Continue pureeing adding small amounts of the bouillon until a soup-like consistency is formed.

Remove the soup to the kettle. Heat slowly. Do not allow the soup to reach the simmer point. Stir frequently. Adjust seasoning.

Garnish: **Garlic Toast**

Remove first layer of skin from an entire bulb of garlic. Place garlic bulb in a baking dish and drizzle a little olive oil over it along with 1/2 teaspoon rosemary. Bake at 350º for 45 minutes. Spread baked garlic cloves on day-old slices of crusty bread.

Basil Tomato Soup

1/2 onion
1 Tbsp. oil of choice (olive, canola, margarine, butter)
4 large, very ripe tomatoes
1 bunch basil
1 medium steamed potato
4 cups vegetable bouillon
1/2 tsp. sugar
1/2 tsp. salt

Mince onion. Heat oil of choice in a kettle and add the onion. Cook until the onion is soft.

Discard the stems of the basil. You should have approximately 1 cup of leaves that are lightly packed.

Quarter the tomatoes and steamed potato.

Alternating basil leaves, tomato and potato, chop in a food processor or blender until the ingredients are assimilated. Add the vegetable bouillon, a little at a time, continuing to work the machine until a soup-like consistency is achieved. Add the seasoning and pour the soup into the kettle.

Warm, stirring, but do not allow the soup to come to a simmer. Adjust seasoning.

Garnish: Cooked noodles

Beet Sauerkraut Soup

1/2 medium onion
2 large cloves garlic
2 Tbsp. oil of choice (olive, canola, margarine, butter)
1 bay leaf
1 cup sauerkraut—juice squeezed out
4 medium beets
4 cups water saved from steaming vegetables or
    fresh water (soup base)
1 medium steamed potato
1 Tbsp. miso (a fermented grain; adds a salty flavor;
    available at health food stores)
Pepper to taste

Chop onion and garlic fine. Crush bay leaf. Heat oil in a soup kettle and add onion and garlic. Sauté until onion is browned. Stir frequently.

Peel and quarter beets. Chop in a blender or food processor a few at a time. Add steamed potato and continue to chop. Add sauerkraut and continue chopping. Add water slowly and continue processing until soup-like consistency is achieved. Pour soup into kettle.

Dissolve miso in 1/2 cup warm water and add to the soup kettle.

Warm the soup slowly to just before it simmers, stirring frequently. Season soup with pepper to taste.

Garnish: Cubes of boiled potatoes and yogurt, soy yogurt or sour cream

Beet Soup Cream Style

1 Tbsp. oil of choice (olive, canola, margarine, butter)
3 Tbsp. oatmeal
4 medium beets
2 cups vegetable bouillon
2 cups milk product of choice (soy milk, evaporated
  skim milk, milk, cream)
1/4 tsp. nutmeg
1/2 tsp. salt

Heat oil of choice in a soup kettle. Add oatmeal and
nutmeg and stir continuously until ingredients are
assimilated. Add milk product of choice.

Peel and quarter beets. Chop in a blender or food
processor using a little of the bouillon to keep the blade
moving.

Add the remaining bouillon, a little at a time, continuing to work the machine until a soup-like consistency is
achieved. Season with salt.

Pour soup into kettle. Warm, stirring, but do not allow
the soup to come to a simmer. Taste and adjust seasoning.

Garnish: Raw green peas

Beet Soup Polish Style

1 Tbsp. oil of choice (olive, canola, margarine, butter)
1/2 small onion
4 medium beets
2 medium apples
2 medium tomatoes
1 medium steamed potato
1/4 small cabbage
4 cups vegetable bouillon
1 tsp. sugar
Salt and pepper to taste

Chop onion fine and add to oil of choice. Heat in a soup pot until onions are soft. Stir frequently.

Quarter beets, apples, tomatoes and potato. Chop beets in a food processor or blender.

Add tomatoes, apples, potato and cabbage a few pieces at a time, continuing to process the machine. Add a little of the bouillon to keep the blade moving. Add the remaining bouillon, pureeing until a soup-like consistency is achieved.

Transfer soup to the pot and warm slowly. Do not allow the soup to come to a simmer. Stir frequently. Add sugar, salt and pepper.

Garnish: Cooked lima beans

Beet Soup Russian Style

1 small onion
1 Tbsp. oil of choice (olive, canola, margarine, butter)
4 medium beets
1 medium steamed potato
1 medium tomato
4 cups vegetable bouillon
1 Tbsp. Spike (seasoning available at health food store)
1 tsp. salt
1 tsp. sugar
Lemon juice to taste

Heat oil of choice in a soup kettle. Mince onion and brown.

Peel and quarter beets. Chop in a food processor or blender. If necessary, use a little of the bouillon to keep the blade moving.

Add the potato and the tomato, continuing to puree until the ingredients are assimilated. While working the machine, combine the seasoning and the remaining bouillon, a little at a time, with the puree until a soup-like consistency is achieved.

Pour the soup into the kettle and warm slowly. Do not allow the soup to reach a simmer. Stir frequently. Add lemon juice to taste.

Garnish: Chopped cucumbers and sour cream. If desired, substitute soy sour cream.

Beet Soup Simple

1 small onion
2 cloves garlic
1 Tbsp. oil of choice (olive, canola, margarine, butter)
4 medium beets
1 small steamed potato
3 cups vegetable bouillon
1 cup buttermilk, or use soy yogurt and
    1/2 Tbsp. lemon juice
2 tsp. salt

Heat oil of choice in a soup pot. Mince onion and garlic and sauté until tender.

Peel beets and quarter. Chop in a food processor or blender. Use a little of the bouillon to keep the blade moving. Add the remaining bouillon and buttermilk, a little at a time. Continue to work the machine until a soup-like consistency is achieved. Add salt. Give the machine another whirl. Taste and adjust the seasoning.

Pour soup into pot. Warm until just before the simmer point. Stir often.

Garnish: Steamed yellow Finn potatoes sprinkled with chopped parsley

Beet Soup Spicy Style

1/2 onion
2 cloves garlic
2 Tbsp. dry basil
2 Tbsp. oil of choice (olive, canola, margarine, butter)
1 Tbsp. dry dill
2 Tbsp. apple cider vinegar
1 Tbsp. honey
4 medium beets
2 medium carrots
1/4 small cabbage
1 red bell pepper
3 medium tomatoes
1 small steamed potato
4-5 cups chicken-flavored bouillon, depending on the
    consistency you like
Salt and pepper to taste

Mince onion and garlic. Heat oil of choice in a large
kettle and sauté the onion and garlic until tender. Add
the basil and dill and cook a few seconds longer, stirring
constantly. Add apple cider vinegar and honey.

Cut beets, carrots, cabbage, red pepper, tomatoes and
steamed potato into 1" chunks.

Chop the vegetables, a few at a time, in a food processor
or blender, alternating the hard vegetables with the
tomatoes and steamed potato. Add small amounts of the
bouillon to keep the blade moving. Add the remaining
bouillon, a little at a time, until a soup-like consistency is
achieved.

Combine the soup with the sauté. Warm until just
below the simmer point. Stir often.

Garnish: Lemon wedges and chopped fresh dill or
parsley

Beet Soup Zesty Style

1/2 medium onion
2 cloves garlic
3 jalapeño peppers
2 Tbsp. oil of choice (olive, canola, margarine, butter)
1/2 tsp. thyme
1/2 tsp. parsley
1/2 tsp. basil
1/2 tsp. marjoram
1/4 tsp. sage
1/4 tsp. rosemary
1/4 tsp. dried, ground lemon peel
4 medium beets
1 carrot
1/2 small cabbage
1 small steamed potato
4-5 cups vegetable bouillon depending on the
    consistency desired or if unexpected company arrives
Salt and pepper to taste

Chop onion, garlic and pepper fine. Heat oil of choice in a kettle and add chopped ingredients. Cook until they are soft. Add all of the herbs and cook a few seconds longer, stirring constantly.

Cut beets, carrot, cabbage and steamed potato into 1" chunks. Chop vegetables in a food processor or blender, alternating chunks of carrots and beets with the cooked potato and cabbage. Add small amounts of the bouillon to keep the blade moving. Add the remaining bouillon and continue chopping until a soup-like consistency is achieved.

Remove soup to kettle. Warm slowly. Do not allow the soup to reach a simmer. Stir frequently. Season to taste with salt and pepper.

Garnish: **Rye Croutons**

Toast slices of day-old rye bread in a 350° oven for several minutes until crisp. Cut into crouton-size pieces.

Broccoli Soup Cream Style

1/2 medium onion
1 Tbsp. oil of choice (olive, canola, margarine, butter)
1 bay leaf
1/4 tsp. nutmeg
1/4 tsp. thyme
2 cups milk product of choice (soy milk, evaporated
    skim milk, milk, cream)
1 lb. broccoli
1 medium steamed potato
2 cups chicken-flavored bouillon
Tamari to taste

Chop onion fine. Heat oil of choice in a kettle. Add
onion, thyme and bay leaf. Sauté until the onion is soft.
Add milk product of choice and warm. Remove the bay
leaf after the milk is warm.

Remove the stems from the broccoli. Save for salads and
stir-frying. Chop the broccoli spears in a food processor
or blender. Quarter the potato and add it to the work
bowl and blend. Add the bouillon, a little at a time,
continuing to puree until a soup-like consistency is
achieved.

Transfer soup to the kettle and warm slowly. Do not
allow the soup to come to a simmer. Stir frequently. Add
tamari to taste.

Garnish: Dust with paprika and add sour cream or soy
sour cream, if desired.

Broccoli Soup
Honey Mustard Flavor

1 Tbsp. oil of choice (butter preferred; olive, canola or
   margarine)
3 Tbsp. honey
1 Tbsp. prepared mustard
1 lb. broccoli
1 medium steamed potato
4 cups chicken-flavored bouillon
Salt and pepper to taste

Melt butter in a soup pot. Add the honey and mustard
and heat thoroughly. Stir frequently to assimilate the
ingredients.

Remove the stems from one pound of broccoli. Save
them for the garnish. Chop broccoli spears in a food
processor or blender. Quarter the potato and add it to the
work bowl. Continue to chop until the ingredients are
assimilated, using small amounts of the bouillon to keep
the blade moving. Add the remaining bouillon, and puree
until a soup-like consistency is achieved.

Transfer the soup to the pot. Warm slowly, but do not
let the soup come to a simmer. Stir frequently. Adjust the
seasoning with salt and pepper.

Garnish: Peeled broccoli stems sliced thin and toasted
sesame seeds. Toast seeds in a dry pan for about
5 minutes. Shake the pan frequently to keep the seeds
from burning.

Broccoli Soup Japanese Style

1 medium onion
4 cloves garlic
2" piece ginger
1-1/2 Tbsp. toasted sesame oil
1 lb. broccoli
1 medium steamed potato
4 cups vegetable bouillon
1 Tbsp. miso
1 Tbsp. tamari

Heat sesame oil in a kettle. Chop onion and garlic fine. Add them to the kettle and cook until onion and garlic are slightly brown. Stir often. Peel ginger. Place in blender with 1 tablespoon water and blend. Add blended ginger to kettle.

Remove broccoli stems. Save for stir-frying and salads. Chop broccoli florets in a food processor or blender. Add steamed potato and continue chopping until the ingredients are assimilated. Use a little of the bouillon to keep the blade moving. Add the remaining bouillon, miso and tamari and continue to puree until a soup-like consistency is achieved.

Transfer soup to kettle. Warm just below the simmer point. Stir often. Adjust the seasoning with additional tamari.

Garnish: **Marinated Tofu Squares**

Cut tofu into 1" squares and marinate in tamari for 1 hour. Stir fry until tofu is firm.

Broccoli Soup Scandinavian Style

1 small onion
1 stalk celery
1/4 tsp. dry mustard
1 Tbsp. oil of choice (olive, canola, margarine, butter)
1 cup milk product of choice (soy milk, evaporated
     skim milk, milk, cream)
1 lb. broccoli
1 medium steamed potato
3 cups chicken-flavored bouillon
Salt and pepper to taste

Chop onion and celery fine. Heat oil of choice in a soup pot. Sauté the onion and celery until they are soft. Stir often. Add milk product of choice and warm.

Remove the stems from the broccoli spears and save for salads or stir-frying. Chop the spears in a food processor or blender. Quarter the potato and add it to the mixture and blend.

Add the bouillon, a little at a time, while continuing to process the machine until a soup-like consistency is achieved.

Transfer soup to the pot and warm slowly. Do not allow the soup to come to a simmer. Stir frequently. Adjust the seasoning with salt and pepper to your taste.

Garnish: Sprigs of fresh dill and grated lemon rind

Broccoli Soup Simple

1 small onion
1 stalk celery
1 Tbsp. oil of choice (olive, canola, margarine, butter)
Dash cayenne
1 tsp. dry basil
1 lb. broccoli
1 medium steamed potato
4 cups chicken-flavored bouillon
1/2 cup milk product of choice (soy yogurt, regular
    yogurt, sour cream)
Salt and pepper to taste

Chop onion and celery fine. Add to oil of choice along with basil and cayenne. Heat in a soup pot until celery and onions are soft. Stir frequently. Add milk product of choice.

Cut the broccoli spears off the stems. Do not discard the stems; when peeled and sliced, they make a crunchy addition to a green salad. Chop the spears in a food processor or blender. Quarter the potatoes and add them to the mixture and blend. Add small amounts of the bouillon, continuing to work the machine until a soup-like consistency is formed.

Transfer soup to the pot and warm slowly. Do not allow the soup to come to a simmer. Stir frequently. Adjust the seasoning with salt and pepper to your taste.

Garnish: Crumbled Roquefort cheese

Brussels Sprouts Soup Spicy Style

1 medium onion
4 cloves garlic
1 Tbsp. oil of choice (olive, canola, margarine, butter)
2" piece ginger
3/4 Tbsp. curry
8 Brussels sprouts
1 large steamed yam
3 cups leftover vegetable cooking water or 1 cup water
1 cup milk product of choice (soy milk, evaporated
    skim milk, milk, cream)
Tamari to taste

Chop onion and garlic fine. Heat oil of choice in a soup kettle and sauté until tender. Peel ginger and shred. Add ginger and curry to sauté and cook for one minute longer. Add milk product of choice and cook until the milk is warmed. Stir well.

Cut Brussels sprouts into 1" pieces. Chop in a food processor or blender. Add steamed yam. Use a little of the leftover vegetable cooking water to keep the blade moving. Add remaining liquid slowly, processing the machine until a soup-like consistency is achieved.

Remove soup to kettle. Warm to serving temperature, but do not allow the soup to simmer. Season to taste with tamari. Stir frequently.

Garnish: Chopped cilantro

Cabbage Tomato Soup

1/2 small onion
1 clove garlic
1 Tbsp. oil of choice (olive, canola, margarine, butter)
1/2 medium-size cabbage
1 medium steamed potato
3 sweet peppers
2 tomatoes
4 cups tomato juice
Salt and pepper to taste

Chop onion, garlic and peppers fine. Combine with oil of choice in a soup pot and cook until soft.

Cut cabbage into 1'' chunks and chop in a food processor or blender. Quarter tomatoes and potato and add them to the work bowl a few pieces at a time. Continue to chop. Add the tomato juice, a little at a time, and puree the ingredients until a soup-like consistency is achieved.

Transfer soup to pot and warm slowly. Do not allow the soup to come to a simmer. Stir frequently. Season to taste with salt and pepper.

Garnish: Spread slices of rye bread with oil of choice and thinly sliced red onion. Place under the broiler for a few minutes.

Cabbage Soup Swedish Style

1 Tbsp. oil of choice (olive, canola, butter, margarine)
1 small onion
2 tsp. caraway seeds
1 cup tomato juice
1/2 small cabbage
1 small steamed potato
3 cups vegetable bouillon
Salt and pepper to taste

Heat oil of choice in a soup pot. Chop onion fine and sauté. Grind caraway seeds in a food mill and add to sauté. Stir well. Add tomato juice and cook a few minutes longer.

Cut cabbage into chunks to accommodate the food processor or blender. Chop until the cabbage is fine. Add steamed potato and process. Add the vegetable bouillon, a little at a time, pureeing until a soup-like consistency is achieved.

Pour mixture into soup pot and warm. Stir frequently. Do not let the soup come to a simmer. Salt and pepper to taste.

Garnish: Whole caraway seeds

Cabbage Soup Sweet-and-Sour

1 Tbsp. oil of choice (olive, canola, margarine, butter
1/2 medium onion
3/4 cup sweet-and-sour sauce (purchase in market)
1/2 small cabbage
1/2 small green pepper
2 small carrots
1 small steamed potato
4 cups vegetable bouillon
Salt and pepper to taste

Heat oil of choice in a kettle. Chop onion fine and sauté until translucent. Add sweet-and-sour sauce and cook one minute longer. Stir well.

Discard core of cabbage. Cut cabbage into 2-inch chunks. Discard green pepper seeds. Cut pepper, carrots and potato into 1'' chunks. Chop cabbage in a food processor or blender. Add pepper, carrots and potato, a few pieces at a time, processing until the ingredients have been assimilated. Use a little of the bouillon to keep the blade moving.

Add the remaining bouillon, continuing to chop until a soup-like consistency is achieved.

Transfer soup to kettle. Warm just below simmer point. Stir frequently. Season to taste with salt and pepper.

Garnish: Small chunks of pineapple (fresh or canned)

Cabbage Soup Zesty Style

1/2 medium onion
1 Tbsp. oil of choice (olive, canola, margarine, butter)
2 tsp. ground caraway seeds (use mill to grind seeds)
1/2 tsp. dry savory
1/2 tsp. dry thyme
1/2 tsp. dry basil
1 medium red bell pepper
1 small cabbage
1 medium steamed potato
4 cups vegetable-flavored bouillon
Salt and pepper to taste

Mince onion. Heat oil of choice in kettle. Add onion, ground caraway seeds, savory, thyme and basil. Cook until the onion is soft. Stir frequently.

Remove the center core from the cabbage. Do not discard. Munch on it while you're cooking! It is crunchy tasting, low in calories, high in fiber and good for you. Cut the cabbage into 1″ chunks and chop in a food processor or blender.

Remove seeds from the red pepper. Do not use them in the soup, but you can sprinkle them into a salad. Quarter the red pepper and steamed potato. Feed the pieces into the machine with small amounts of the bouillon, continuing to chop until the ingredients are assimilated.

Add the remaining bouillon, processing until a soup-like consistency is achieved.

Pour the soup into the kettle and warm slowly until just below the simmer point. Stir often.

Garnish: **Rye Croutons**

Toast several-day-old rye bread in the toaster. Cut into 1/2-inch chunks.

Brussels Sprouts Dahl Soup

1 cup dahl (red lentils)
4 cups water
1 bay leaf
1-1/2 Tbsp. olive oil
4 cloves garlic
1 medium onion
2" piece ginger
1 tsp. turmeric
2 Tbsp. tomato paste
2 tsp. cumin seeds
10 Brussels sprouts
2 medium, very ripe tomatoes
Tamari to taste

Bring water to boil. Add dahl and bay leaf. Lower heat and cook for 1/2 hour until dahl is soft. Discard bay leaf. Puree cooked dahl in a blender. Allow to cool. This will be your soup base.

Heat olive oil in a soup pot. Peel ginger. Chop onion, garlic and ginger fine. Sauté until tender. Add turmeric and tomato paste. Stir well.

In a dry frying pan, roast cumin seeds over high heat for a few minutes. Grind roasted cumin seeds in a food mill and add to the sauté. Cook one minute longer. Stir thoroughly.

Discard the tough stem portion of the Brussels sprouts. Cut sprouts in half and chop in a food processor or blender. Add tomatoes to the work bowl and process. Add dahl soup base, a little at a time, chopping until a soup-like consistency is achieved.

Remove soup to pot. Warm just below simmer point. Stir often. Season to taste with tamari.

Garnish: Grated lemon rind

Carrot Chowder

1/2 small onion
1 Tbsp. oil of choice (olive, canola, margarine, butter)
1/4 tsp. paprika
1/2 tsp. thyme
1-1/2 cups milk product of choice (soy milk,
    evaporated skim milk, milk, cream)
6 medium carrots
1 medium steamed potato
4 cups chicken-flavored bouillon
Salt and pepper to taste

Chop onion fine. Place in soup kettle with oil of choice, paprika and thyme. Cook until onions are soft. Stir frequently. Add milk product of choice.

Cut carrots into 1'' chunks. Chop in a food processor or blender. Quarter potato and add it to the chopped carrots, continuing to work the machine. Use small amounts of the bouillon to keep the blade moving. Add the remaining bouillon, and puree until a soup-like consistency is formed.

Pour the soup into the kettle and warm slowly. Do not allow the soup to come to a simmer. Stir frequently. Adjust the seasoning with salt and pepper.

Garnish: Chopped chives

Carrot Coconut Soup

1 Tbsp. butter or margarine
4 Tbsp. uncooked oatmeal
1 Tbsp. ginger powder
5 medium carrots
4 cups coconut milk (can be purchased frozen or
    canned)
1 cup water
Salt

Heat butter or margarine in a soup pot. Add oatmeal and ginger. Cook over a low flame for 2-3 minutes, stirring constantly.

Cut carrots into 1" pieces and chop in a food processor or blender. Add the coconut milk and water, a little at a time, continuing to process the machine until a soup-like consistency is achieved.

Transfer to soup pot. Warm slowly just below the simmer point. Stir well.

Garnish: Small chunks of fresh pineapple sprinkled with chopped mint.

Carrot Dill Soup

1/2 large onion
2 Tbsp. oil of choice (olive, canola, margarine, butter)
4 Tbsp. dried dill
1/4 cup white wine
5 medium carrots
1 medium steamed potato
5 cups chicken-flavored bouillon
1/2 cup ricotta or soy cream cheese
Salt and pepper to taste

Chop onion fine. Heat oil in a soup pot and add chopped onion. Sauté until the onions are golden brown. Add white wine and dill. Stir well.

Cut carrots into 2″ chunks. Place carrot chunks a few at a time in the work bowl of the food processor or blender and chop fine. Add steamed potato and process. Use a little of the bouillon to keep the blade moving. Add the remaining bouillon, processing until a soup-like consistency is achieved.

Transfer soup to pot. Warm to serving temperature, but do not let the soup reach a simmer. Add ricotta cheese or soy cream cheese. Stir well. Season to taste with salt and pepper.

Garnish: Chopped fresh dill

Carrot Oatmeal Soup

1 Tbsp. oil of choice (olive, canola, margarine, butter)
1/2 small onion
1 tsp. dry dill
1/2 cup oatmeal
1 cup boiling water
4 medium carrots
3 cups vegetable bouillon
1 cup buttermilk (or 1 cup soy milk and
    3/4 Tbsp. lemon juice)
1 tsp. sugar
Salt and pepper to taste

Heat oil of choice in a soup pot. Chop onion fine. Add onion and dill to oil of choice and sauté until onion is soft. Stir often.

Add boiling water to oatmeal and stir well. Do not cook. Cut carrots into 1″ pieces and chop in a food processor or blender. Add oatmeal and continue chopping until the ingredients are assimilated. Use the remaining bouillon and buttermilk, a little at a time, working the machine until a soup-like consistency is achieved.

Transfer soup to the pot. Warm slowly to just below the simmer point. Stir well. Add sugar and season to taste with salt and pepper.

Garnish: Fresh green peas

Carrot Soup Curry Flavor

1 Tbsp. oil of choice (olive, canola, margarine, butter)
1 medium onion
1/2 tsp. cumin
1/2 tsp. curry powder
1 Tbsp. lemon juice
5 medium carrots
1 medium steamed potato
4 cups chicken-flavored bouillon
Salt and pepper to taste

Chop onion fine. Heat oil of choice in a soup kettle and sauté onion until transparent. Add cumin, curry and lemon juice. Stir well.

Cut carrots into 1" pieces and chop in a blender or food processor. Add steamed potato. Use a little of the bouillon to keep the blade moving. Add the remaining bouillon, pureeing until a soup-like consistency is formed.

Pour soup into kettle. Warm to serving temperature, but do not allow the soup to reach a simmer. Stir often. Season to taste with salt and pepper.

Garnish: Blend 1 bunch watercress and 1/2 cup yogurt or soy sour cream.

Carrot Peanut Butter Soup

4 medium carrots
4 cups chicken-flavored bouillon
1/2 cup peanut butter
1/4 tsp. asafoetida (available in health food stores)
1 Tbsp. curry powder
Salt and pepper to taste

Cut carrots into 1″ chunks. Chop in food processor or blender until fine. Add peanut butter, curry powder and asafoetida. Continue to work the machine while adding small amounts of the bouillon until a soup-like consistency is formed.

Pour the mixture into a soup pot and warm. Do not allow soup to simmer. Stir frequently. Season with salt and pepper to taste.

Garnish: Cooked brown rice; raw broccoli stems, sliced thin

Carrot Sunflower Seed Soup

5 medium carrots
3/4 cup sunflower seeds
3 cups chicken-flavored bouillon
1 cup orange juice
Salt and pepper to taste

Soak sunflower seeds in water to cover, overnight.

Cut carrots into 1" pieces. Chop in food processor or blender. Drain sunflower seeds. Discard soaking water. Add seeds to work bowl and process. Use a little of the bouillon to keep the blade moving.

Add the remaining bouillon and orange juice, a little at a time, continuing to chop until a soup-like consistency is achieved.

Remove soup to a kettle. Warm slowly. Do not allow the soup to reach a simmer. Season to taste with salt and pepper.

Garnish: Raisins and, if desired, sour cream

Carrot Soup Orange Flavor

1 small onion
1 Tbsp. oil of choice (olive, canola, butter, margarine)
1/4 tsp. powdered cloves
2 cups fresh orange juice
Juice of 1/2 lemon
5 medium carrots
1 medium steamed potato
3 cups leftover vegetable cooking water or
    use plain water
Salt and pepper

Chop onion fine. Heat oil of choice in a soup pot. Sauté onions until soft. Add cloves, sherry, orange juice and lemon juice. Stir well.

Cut carrots into 1'' chunks and chop in a food processor or blender. Add the potato and process. Add leftover vegetable cooking water, a little at a time, working the machine until a soup-like consistency is achieved.

Transfer soup to pot. Warm until serving temperature. Do not allow the soup to come to a simmer. Stir frequently. Season to taste with salt and pepper.

Garnish: Grated orange rind

Cauliflower Chowder

1 small onion
3 cloves garlic
1 tsp. ground caraway seeds
1/2 tsp. dry mustard
1 Tbsp. oil of choice (olive, canola, margarine, butter)
1 small cauliflower
2 small carrots
1 steamed potato
4 cups vegetable bouillon
Salt and pepper to taste

Chop onion and garlic fine. Grind caraway seeds in a food mill. Heat oil of choice in a kettle. Add onion, garlic, caraway seed and mustard. Sauté until the onion and garlic are soft. Stir frequently.

Remove the core from the cauliflower and discard. Chop florets in a food processor or blender. Cut carrots in 1" pieces. Quarter potato. Add carrots and potato to the machine a few at a time and continue chopping until the ingredients are assimilated. Use small amounts of the bouillon to keep the blade moving. Add the remaining bouillon, working the machine until a soup-like consistency is achieved.

Transfer the soup to the kettle. Warm just below the simmer point. Stir frequently. Season to taste with salt and pepper.

Garnish: Chopped scallions and grated Cheddar cheese. If desired, soy Cheddar may be used.

Cauliflower Coconut Soup

1 Tbsp. oil of choice (olive, canola, margarine, butter)
1/2 small onion
1-1/2 fennel seeds
2 cups coconut milk (can be purchased canned
    or frozen)
1 small cauliflower
1 small steamed potato
2 cups leftover vegetable cooking water (or plain water)
Salt to taste

Heat oil of choice in a soup pot. Chop onion fine and sauté until transparent. In a dry frying pan, toast fennel seeds for a few minutes over high heat. Grind them in a food mill. Combine the ground fennel seeds and coconut milk with the sauté. Cook one minute longer. Stir well.

Remove the core and any tough stem portions from the cauliflower. Chop the cauliflower florets in a food processor or blender. Add the steamed potato and process until the ingredients are assimilated. Add the vegetable cooking water, a little at a time, pureeing until a soup-like consistency is achieved.

Transfer soup to pot. Warm slowly to serving temperature. Do not allow the soup to reach the simmer point. Stir frequently. Adjust seasoning with salt.

Garnish: Fresh shredded coconut

Cauliflower Soup Cream Style

1 Tbsp. margarine or butter
1 small onion
2 tsp. dried dill
1 tsp. thyme
1 cup milk product of choice (soy milk, evaporated
   skim milk, milk, cream)
2 bay leaves
1 small cauliflower
1 medium steamed potato
3 cups chicken-flavored bouillon
Salt and pepper to taste

Heat margarine or butter in a kettle. Chop onion fine
and sauté until transparent. Grind bay leaves in a food
mill. Add ground bay leaves, dill and thyme to the sauté.
Stir well. Add milk product of choice and cook one
minute longer.

Remove the core and any tough stem portions from the
cauliflower. Chop the florets in a food processor or
blender. Add the steamed potato to the work bowl,
continuing to process until the ingredients are assimilated. Add the chicken-flavored bouillon, a little at a time,
pureeing until a soup-like consistency is achieved.

Transfer soup to kettle. Warm just below the simmer
point. Stir frequently. Season to taste with salt and
pepper.

Garnish: Fresh dill

Cauliflower Soup Caribbean Style

1 medium onion
2 jalapeño peppers
2 Tbsp. dry tarragon
1-1/2 Tbsp. oil of choice (olive, canola, margarine,
    butter)
1 small cauliflower
1 medium red pepper
1 small steamed potato
4 cups chicken-flavored bouillon
Salt and pepper to taste

Chop onion fine. Slice peppers thin. Combine with
tarragon and oil of choice in a soup kettle and cook until
the onions are golden brown. Stir often.

Discard the leaves and stem of the cauliflower. Separate
the spears and chop in a food processor or blender.
Discard the seeds from the red pepper and cut it into
quarters. Add this to the machine and continue chopping.
You may need a little of the bouillon to keep the blade
moving. When the ingredients are assimilated, add the
steamed potato with a little of the bouillon, continuing to
work the machine. Add the remaining bouillon, and
puree the ingredients until a soup-like consistency is
formed.

Pour the soup into the kettle and warm slowly. Do not
allow the soup to come to a simmer. Stir frequently.
Season to taste with salt and pepper.

Garnish: Chopped parsley

Cauliflower Soup Thai Style

4 heaping Tbsp. dried lemon grass
    (can be purchased in a health food store)
4 cups boiling water
1 medium onion
4 cloves garlic
1 Tbsp. olive oil or canola oil
1 tsp. sweet basil
1 Tbsp. curry powder
3 Tbsp. tomato paste
1 small cauliflower
1 small steamed potato
Tamari to taste

Add boiling water to lemon grass and allow to steep for
half an hour. Strain and discard leaves. The lemon grass
tea will be your soup base.

Chop onion and garlic fine. Heat olive oil or canola oil
in a soup kettle. Sauté onion and garlic until tender. Add
sweet basil, curry powder and tomato paste. Cook one
minute longer. Stir well.

Separate cauliflower florets from core. Do not use the
core. However, it can be saved for stir-frying or salads.
Chop the cauliflower florets in a food processor or
blender. Add potato and process. Use a little of the lemon
grass soup base to keep the blade moving. Add the
remaining soup base slowly, continuing to chop until a
soup-like consistency is achieved.

Transfer soup to kettle. Warm to serving temperature,
but do not allow the soup to simmer. Stir frequently.
Season to taste with tamari.

Garnish: Thinly sliced zucchini

Cauliflower Soup Spicy Style

1 clove minced garlic
1 small, chopped onion
1 Tbsp. oil of choice (olive, canola, margarine, butter)
1/2 tsp. dry mustard
1/2 tsp. ginger
1/2 tsp. cumin
1/2 tsp. turmeric
Pinch cinnamon
Pinch cayenne pepper
1 Tbsp. honey
1/2 Tbsp. lemon juice
5 medium carrots
1 small steamed potato
2 cups milk product of choice (soy milk, evaporated
    skim milk, milk, cream)
2 cups vegetable bouillon
Salt

Chop onion and garlic fine. Heat oil in a soup pot. Sauté onion and garlic until they are golden. Add mustard seed, ginger, cumin, turmeric, cinnamon, cayenne, honey and lemon juice. Cook one minute longer. Stir well.

Cut carrots into 2" chunks. Chop the carrots a few at a time in the food processor or blender. Add a little of the bouillon, if necessary, to keep the blade moving. Add the potato and continue chopping. Add the bouillon, slowly blending until the ingredients are pureed. Add milk product of choice and continue pureeing until a soup-like consistency has been achieved.

Empty work bowl into soup pot. Warm soup to serving temperature, but do not allow the soup to simmer. Stir frequently. Adjust seasoning with salt.

Garnish: Raw peas

Celery Barley Soup

1/2 small, chopped onion
1 Tbsp. oil of choice (olive, canola, margarine, butter)
1/4 tsp. ground cloves
1/4 tsp. thyme
3 cups vegetable bouillon
1/2 cup cooked barley
1 cup milk product of choice (soy milk, evaporated skim
    milk, skim milk, milk, cream)
7 stalks celery (include leaves)
Salt and pepper to taste

In a soup kettle, add chopped onion, cloves and thyme
to oil. Sauté until onion is transparent. Stir frequently.

Quarter celery and puree in blender or food processor.
Add vegetable bouillon, and continuing pureeing. Add
this mixture to the sauté. Add cooked barley and milk
product to the soup kettle.

Warm the soup slowly to just below the simmer point,
stirring frequently. Season with salt and pepper to taste.

Garnish: Chopped parsley

Celery Soup Simple

1 Tbsp. oil of choice (olive, canola, butter, margarine)
1 clove garlic
1/2 small onion
7 stalks celery
1 medium steamed potato
4 cups vegetable bouillon
1 Tbsp. Spike, or more if desired (Spike available at
    health food store)

Heat oil of choice in a soup pot. Chop onions and garlic
fine and add to soup pot. Sauté until onion is soft.

Trim celery and save the leaves for garnish. Cut celery
into 1″ pieces. Chop the celery in a food processor or
blender until fine. Add the steamed potato and continue
chopping. Add the bouillon, a little at a time, continuing
to work the machine until a soup-like consistency is
achieved. Add Spike.

Combine this mixture with the ingredients in the soup
pot and warm. Do not allow the soup to come to a
simmer. Stir frequently.

Garnish: Celery leaves chopped fine

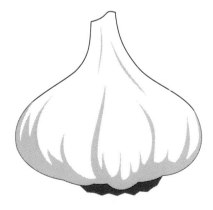

Celery Soup Cream Style

1/2 small onion
1 Tbsp. oil of choice (olive, canola, butter, margarine)
1 cup milk product of choice (soy milk, evaporated skim
    milk, milk, cream)
1/4 tsp. cloves
1 bay leaf
1/4 tsp. thyme
7 stalks celery (include leaves)
1 medium steamed potato
3 cups vegetable bouillon
Salt to taste

Chop onion fine. Heat oil of choice in a soup pot and
add the onion. Sauté until the onion is soft.

Add milk product of choice, bay leaf, cloves and thyme
and keep warm on a small flame until the rest of the
ingredients are ready. Stir from time to time. This allows
the milk product to take on the flavor of the spices.

Cut celery into 1" pieces. Place celery and leaves into a
food processor or blender and chop until fine. Add the
steamed potato and continue chopping. Add the bouillon,
a little at a time, continuing to work the machine until a
soup-like consistency is achieved.

Add soup to pot and warm. Do not allow the soup to
come to a simmer. Stir frequently. Salt to taste.

Garnish: Grated hard-boiled egg

Celery Tomato Soup

2 cups milk product of choice (soy milk, evaporated
  skim milk, milk, cream)
1 bay leaf
6 thin slices onion
1 tsp. paprika
2 Tbsp. tomato paste
1 bunch celery
2 medium tomatoes
1 medium steamed potato
2 cups vegetable bouillon
Salt and pepper to taste

Heat milk product of choice slowly in a soup pot. Add
bay leaf, onion, paprika and tomato paste. Cook until the
onion is tender. Stir frequently. Allow to cool and discard
the bay leaf.

Cut celery stalks into 1" pieces. Chop stalks and leaves
in a food processor or blender. Quarter tomatoes and
potato. Alternating pieces, combine them with the
chopped celery, working the machine until the ingredients are assimilated. Use small amounts of the bouillon
to keep the blade moving. Add the remaining bouillon,
and puree until a soup-like consistency is achieved.

Transfer soup to pot and warm slowly. Do not allow the
soup to reach a simmer. Stir frequently. Season to taste
with salt and pepper.

Garnish: Make an omelet large enough to give each
diner several slices. Allow the omelet to cool; slice thinly.

Chard Soup Cream Style

1/2 medium onion
1 stalk celery
1 Tbsp. oil of choice (olive, canola, margarine, butter)
1 lb. chard
1 medium steamed potato
2 cups buttermilk—or if desired, soy milk can be used
    with 3/4 Tbsp. lemon juice
2 cups chicken-flavored bouillon
Salt and pepper to taste

Chop onion and celery fine. Add to oil of choice and heat in a soup pot until they are soft. Stir frequently. Add buttermilk or soy milk mixture. Cook one minute longer.

Clean chard and discard the tough ends of the stems. Chop chard leaves in a food processor or blender. Quarter steamed potato and add to the chard, continuing to chop until the ingredients are assimilated. Add the bouillon, a little at a time to the work bowl and puree until a soup-like consistency is formed.

Transfer soup to the pot. Warm slowly. Do not allow the soup to reach a simmer. Stir frequently. Add salt and pepper to taste.

Garnish: Small thin-skinned, steamed potatoes, quartered; and chopped cucumber

Chard Soup Curry Flavor

4 shallots
1 Tbsp. oil of choice (olive, canola, margarine, butter)
2 tsp. curry powder
1/2 tsp. cumin
1/2 tsp. coriander
1 cup milk product of choice (soy milk, evaporated
    skim milk, milk, cream)
1 bunch chard
1 medium steamed potato
3 cups chicken-flavored bouillon
Salt to taste

Mince shallots and sauté them in oil of choice. Add
curry powder, cumin, coriander and milk product of
choice. Cook until milk is warm. Stir well.

Remove tough ends of chard and discard. Tear chard in
1″ pieces and chop in a food processor or blender. Add
potato, chopping until assimilated with chard. Use a little
of the bouillon to keep the blade moving. Add the
remaining bouillon, pureeing until a soup-like consistency
is achieved.

Transfer soup to a pot and warm slowly over low flame
to just below the simmer point. Adjust seasoning with
salt.

Garnish: Cooked lentils

Chard Soup Egyptian Style

1 clove elephant garlic
1 Tbsp. oil of choice (olive, canola, margarine, butter)
1 lb. chard
1 medium cucumber
1 small steamed potato
4 cups vegetable bouillon
1/2 cup soy yogurt or dairy yogurt
1 tsp. sugar
2 tsp. lemon juice
Salt and pepper to taste

Crush garlic in a garlic press. Heat oil of choice in a soup pot. Add the garlic and sauté until the garlic is lightly brown. Stir often.

Remove the tough stems from the chard and discard them. Chop the leaves in a food processor or blender. Peel the cucumber and cut it into 1" pieces. Quarter the potato. Add the cucumber and potato to the work bowl, chopping until the ingredients are well blended. Use a little of the bouillon to keep the blade moving. Add the remaining bouillon slowly, continuing to process until a soup-like consistency is achieved. Add the soy yogurt or dairy yogurt, sugar and lemon juice. Whirl the machine once again.

Transfer soup to pot. Warm slowly to just below the simmer point. Stir often. Season to taste with salt, pepper.

Garnish: **Miniature pitas stuffed with eggplant**

Dribble a little olive oil on thin slices of egplant. If desired, season with salt and pepper. However, the olive oil itself will impart a salty taste.

Broil the slices for several minutes until they are soft when punctured with a fork. Stuff the miniature pitas with broiled eggplant and cut them into quarters.

Chard Soup Japanese Style

1/2 small onion
1 Tbsp. toasted sesame oil
3 Tbsp. sesame seeds
1 Tbsp. honey
3 Tbsp. hot water
1 small steamed potato
4 cups chicken-flavored bouillon
Tamari to taste

Chop onion fine. Heat sesame oil in a soup kettle. Sauté onion until transparent. In a dry skillet, toast sesame seeds, stirring constantly for a few minutes. Do not let seeds burn. Transfer seeds to a food mill and grind. Add ground seeds to soup kettle with honey and hot water. Cook a few minutes longer until ingredients are well blended.

Remove the tough stems from chard and discard. Chop chard in a food processor or blender. Add potato and process until the ingredients are assimilated. Use a little of the bouillon to keep the blade moving. Add the remaining bouillon, a little at a time, working the machine until a soup-like consistency is achieved.

Transfer soup to kettle. Warm just below simmer point. Stir frequently. Season to taste with tamari.

Garnish: Cooked Soboda noodles (Japanese noodles)

Chard Soup Jewish Style

1 medium onion
3 cloves garlic
1 Tbsp. oil of choice (olive, canola, margarine, butter)
4 cups vegetable bouillon
2 eggs
1 lb. chard
1 medium steamed potato
Salt and pepper to taste

Chop onion and garlic fine. Heat oil of choice in a kettle. Sauté onions and garlic until lightly brown. Stir often. Add one cup vegetable bouillon and simmer five minutes. Beat eggs and add to hot bouillon. Cook a few minutes longer and allow to cool.

Remove thick stems from chard. Chop chard leaves in a food processor or blender. Add the steamed potato and continue to chop until the ingredients are assimilated. Add remaining two cups of bouillon, a little at a time, continuing to process the machine until a puree is formed.

Transfer the puree to kettle and warm just below the simmer point. Stir often. Season to taste with salt and pepper.

Garnish: Thin-skinned, boiled potatoes and sour cream or soy sour cream, if desired.

Chard Soup Simple

1/2 medium onion
1 Tbsp. oil of choice (olive, canola, margarine, butter)
2 Tbsp. caraway seeds
1 lb. chard
1 medium steamed potato
4 cups chicken-flavored bouillon
Spike to taste (seasoning available in health food store)

Chop onion fine. Grind caraway seeds in a food mill. Heat oil of choice in a kettle. Add onion and ground caraway seeds. Sauté until the onion is slightly brown. Stir well.

Remove the tough stems from the chard and discard. Chop the chard in a food processor or blender. Quarter the potato and add it to the work bowl, processing the machine until the ingredients are well blended. Use a little of the bouillon to keep the blade moving. Add the remaining bouillon, a little at a time, pureeing until a soup-like consistency is achieved.

Transfer soup to kettle. Warm slowly to just below the simmer point. Stir frequently. Season to taste with Spike.

Garnish: Whole caraway seeds

Chard Tomato Soup

1 Tbsp. oil of choice (olive, canola, margarine, butter)
1 medium onion
3 cloves garlic
1 tsp. rosemary
3 Tbsp. tomato paste
1 lb. chard
7 fresh basil leaves
4 medium tomatoes
1 small steamed potato
4 cups chicken-flavored bouillon
Salt and pepper to taste

Chop onion and garlic fine. Heat oil of choice in a kettle. Sauté onion and garlic until slightly brown. Add rosemary and tomato paste. Cook one minute longer. Stir well.

Remove thin stems from chard. If you are into juicing, save them and put in your juicer with other vegetables. Chop chard leaves in a food processor or blender. Add basil leaves, tomatoes, steamed potato and continue chopping until the ingredients are assimilated. Add small amounts of bouillon to keep the blade moving. Add the remaining bouillon and continue working the machine until a soup-like consistency is achieved.

Transfer soup to the kettle. Warm to just below the simmer point. Stir often. Season to taste with salt and pepper.

Garnish: **Polenta Squares**

To 4 cups boiling water add 1 cup coarse cornmeal. Stir constantly until the mixture is as thick as porridge. If desired, 1 cup Parmesan or soy Parmesan can be added while polenta is hot. Remove polenta to an oiled baking pan. Allow to cool and cut into squares.

Corn Basil Soup

1 small onion
2 cloves garlic
1 Tbsp. olive oil
1 bunch basil
1/2 cup olive oil
4 ears corn
1 small steamed potato
4 cups chicken-flavored bouillon
Salt and pepper to taste

Heat one tablespoon olive oil in a soup pot. Chop onion and garlic fine. Sauté until transparent. Stir well.

Remove basil leaves and discard stems. Combine basil leaves with half cup olive oil and chop in a food processor or blender until the ingredients are merged.

With a sharp knife, remove kernels from corn and discard cobs. Chop kernels in a food processor or blender. Add potato and process. Add the bouillon, a little at a time, pureeing until a soup-like consistency is achieved.

Transfer soup to pot. Warm just below simmer point. Add 4 heaping tablespoons of the basil mixture. Reserve the rest for baked potatoes or pasta. Season to taste with salt and pepper.

Garnish: Toasted pine nuts

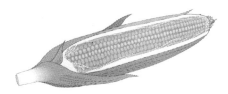

Corn Soup Confetti Style

1 medium red onion
4 cloves garlic
1-1/2 Tbsp. oil of choice (olive, canola, margarine,
    butter)
1 bay leaf
1/4 tsp. thyme
1/2 tsp. basil
3 large ears of corn
2 celery stalks
1/4 cup oatmeal
3 cups vegetable bouillon
1 cup buttermilk (or soy milk with 3/4 Tbsp. lemon
    juice
Salt and pepper to taste

Heat oil of choice in a soup pot. Chop onion and garlic
fine. Crush bay leaf in a food mill. Add onion, garlic,
crushed bay leaf, thyme and basil to the soup pot. Sauté
until the onion and garlic are lightly browned. Stir often.

Remove the kernels from the corn and discard the cobs.
Cut the celery stalks into 1" pieces. Chop the corn kernels
and celery in a food processor or blender. Cover the
oatmeal with hot water and let stand for a few minutes
until the oatmeal is soft. Add the oatmeal to the work
bowl and process. Use a little of the bouillon to keep the
blade moving. Slowly add the remaining bouillon and
buttermilk, or soy milk mixture, continuing to work the
machine until a soup-like consistency is achieved.

Remove soup to pot. Warm just below the simmer
point. Stir often. Season to taste with salt and pepper.

Garnish: Diced green and red peppers

Corn Soup Cream Style

1 Tbsp. oil of choice (olive, canola, margarine, butter)
1 small onion
1/2 tsp. cumin
4 large ears of corn
2 cups milk product of choice (soy milk, evaporated
    skim milk, milk, cream)
2 cups vegetable bouillon
Salt and pepper to taste

Heat oil of choice in a soup kettle. Chop onion fine and
sauté until transparent. Add cumin and milk product of
choice. Cook one minute longer. Stir well.

Cut kernels off cob with a sharp knife. Puree in a food
processor or blender. Add bouillon, a little at a time,
continuing to process until a soup-like consistency is
formed.

Transfer soup to kettle. Warm slowly, stirring
frequently. Do not allow soup to simmer. Season with salt
and pepper to taste.

Garnish: Corn chips

Corn Soup Simple

1 small onion
4 cloves garlic
1 Anaheim pepper
1-1/2 Tbsp. oil of choice (olive, canola, margarine,
   butter)
4 ears of corn
4 cups chicken-flavored bouillon
4 Tbsp. polenta (coarse cornmeal)
Salt and pepper to taste

Chop onion, garlic and pepper fine. Heat oil of choice in
a kettle and sauté the ingredients until they are soft. Stir
well.

Add the bouillon and polenta and bring to a boil. Cook
for several minutes. Stir often. Allow to cool to serving
temperature.

With a sharp knife, remove kernels from the corn.
Discard the cobs. Add corn kernels to kettle.

Season to taste with salt and pepper.

Garnish: Grated Monterey Jack or soy Monterey Jack

Corn Soup Zesty Style

1 Tbsp. oil of choice (olive, canola, margarine, butter)
1 medium onion
3 cloves garlic
3 large ears of corn
2 stalks celery
1 small steamed potato
3 cups vegetable bouillon
1 cup buttermilk (or soy milk with 3/4 Tbsp. lemon
   juice)
1/2 cup salsa (mild or hot) to your taste

Heat oil of choice in a kettle. Chop onion and garlic fine. Sauté until soft. Stir often.

Remove the kernels from three large ears of corn. Discard the cobs (but first, have you ever tasted that sweet "nectar" that remains in the cob after the kernels have been removed?) Chop the corn kernels in a food processor or blender. Cut the celery into 1" pieces. Quarter the potato. Add the celery and potato to the work bowl, continuing to chop. Use a little of the bouillon to keep the blade moving. Add the buttermilk, or soy milk mixture, and salsa, processing the machine. Add the remaining bouillon slowly, pureeing until a soup-like consistency is achieved.

Remove the soup to the kettle. Warm just below the simmer point. Stir frequently. Adjust the seasoning with additional salsa, if desired.

Garnish: Cilantro sprigs

Corn Tomato Soup

4 shallots
1 Tbsp. oil of choice (olive, canola, margarine, butter)
1 tsp. oregano
1/2 tsp. thyme
3 Tbsp. tomato paste
1/4 cup red wine
4 medium ears of corn
2 medium very ripe tomatoes
1 small steamed potato
5 sprigs parsley
4 cups leftover vegetable cooking water or plain water
Salt and pepper to taste

Heat oil of choice in a kettle. Mince shallots. Sauté until lightly brown. Add oregano, thyme, tomato paste and wine. Cook one minute longer. Stir well.

With a sharp knife, remove corn kernels and discard the cobs. Chop the kernels in a food processor or blender. Quarter tomatoes and potato and add them to the work bowl along with parsley leaves (discard stems). Continue chopping until the ingredients are assimilated, using a little of the vegetable cooking water to keep the blade moving. Add the remaining vegetable cooking water slowly, pureeing until a soup-like consistency is achieved.

Transfer to soup kettle. Warm to serving temperature, but do not allow the soup to simmer. Stir frequently. Season to taste with salt and pepper.

Garnish: Thin slices of avocado

Cucumber Dill Soup

3 large cucumbers
8 scallions
1 Tbsp. dry dill
1 medium steamed potato
3 cups chicken-flavored bouillon
1 tsp. balsamic vinegar
1 cup milk product of choice (soy milk, evaporated
    skim milk, or milk)
Salt and pepper to taste

Peel cucumber and cut into 1" chunks. Remove most of
the green end of the scallions. Cut them into 1" pieces.
Chop the scallions and cucumbers in a food processor or
blender. Add the potato, continuing to chop. Add milk
product of choice and bouillon, a little at a time, pureeing
until a soup-like consistency is achieved.

Remove the soup to a kettle. Warm over low heat,
stirring frequently. Do not let soup reach the simmer
point. Add balsamic vinegar and season to taste with salt
and pepper.

Garnish: Fresh dill

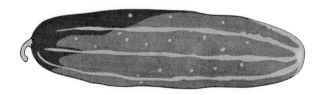

Cucumber Soup Simple

1 small onion
1 Tbsp. margarine or butter
4 Tbsp. oatmeal
1/4 cup hot water
2 Tbsp. honey
2 Tbsp. lemon juice
3 large cucumbers
3 cups vegetable bouillon
1 cup buttermilk or soy milk
Salt and pepper to taste

Chop onion fine. Heat margarine or butter in a soup pot. Sauté onion until translucent. Add oatmeal and hot water. Stir well. Add honey and lemon juice. Cook for one minute longer.

Peel cucumber and cut into 1" chunks. Chop cucumber in a food processor or blender. Add bouillon and buttermilk or soy milk, continuing to process until a soup-like consistency is achieved.

Remove soup to pot. Warm slowly just below the simmer point. Season to taste with salt and pepper.

Garnish: Garlic chives sprinkled with paprika

Cucumber Beet Soup

2 cloves garlic
1/2 Tbsp. oil of choice (olive, canola, margarine, butter)
1 medium beet
8 sprigs parsley
2 green onions
1 medium steamed potato
2 cups milk product of choice (soy milk, evaporated
    skim milk, milk, cream)
2 medium cucumbers
2 cups vegetable bouillon
Salt and pepper to taste

Mince garlic in a garlic press. Heat oil of choice in a
kettle and sauté garlic until slightly brown. Stir often.

Quarter beets and potato. Peel cucumber and cut into
1" pieces. Remove the stems from the parsley and
discard. Cut green onions into 1" pieces. Chop the beets
in a food processor or blender. Add the cucumber,
parsley leaves, green onions and potato a few pieces at a
time, continuing to chop until the ingredients are
assimilated. Use a little of the bouillon to keep the blade
moving. Add the remaining bouillon and milk product of
choice, continuing to work the machine until a soup-like
consistency is achieved.

Transfer soup to the kettle. Warm slowly over low heat
to just below the simmer point. Stir frequently. Season to
taste with salt and pepper.

Garnish: Chopped chives and, if desired, sour cream or
soy sour cream.

Cucumber Tomato Soup

1 small onion
1 Tbsp. oil of choice (olive, canola, margarine, butter)
1 tsp. oregano
1/2 tsp. thyme
4 large cucumbers
3 medium tomatoes
1 small steamed potato
4 cups chicken-flavored bouillon
Salt and pepper to taste

Chop onion fine. Heat oil of choice in a soup pot and sauté onion until translucent. Add oregano and thyme. Cook for one minute longer. Stir well.

Peel cucumbers and cut into 1" chunks. Quarter tomatoes and potato. Chop cucumbers, tomatoes and potato in a food processor or blender. Use a little of the bouillon to keep the blade moving. Add the remaining bouillon, pureeing until a soup-like consistency is achieved.

Transfer soup to pot. Warm just below the simmer point. Season to taste with salt and pepper.

Garnish: Small chunks of cucumber

Fennel Soup Simple

1 Tbsp. oil of choice (olive, canola, margarine, butter)
3 cloves garlic
2 large fennel bulbs
1 medium steamed potato
4 cups chicken-flavored bouillon
Salt and pepper to taste

Press garlic in a garlic press. Heat oil of choice in a
kettle and sauté garlic until lightly brown.

Remove tough core of fennel and leafy tops. Discard
core but save leafy tops for garnish or salad. Cut fennel
bulbs and stalks into 1" chunks. Chop in a food processor
or blender. Add steamed potato and process. Add
bouillon, a little at a time, continuing to work the machine
until a soup-like consistency is formed.

Pour soup into kettle. Warm to serving temperature but
do not allow the soup to simmer. Stir frequently. Season
to taste with salt and pepper.

Garnish: **Garlic Toast**

Toast slices of day-old whole wheat bread. Drizzle a
little olive oil on each slice of toast and rub with garlic.

Fennel Soup Italian Style

1/2 small onion
2 cloves garlic
1 Tbsp. olive oil
1 tsp. basil
1/2 tsp. oregano
3 Tbsp. tomato paste
2 medium fennel bulbs
1 medium steamed potato
1 small tomato
4 cups leftover vegetable cooking water or plain water
Salt and pepper to taste

Chop onion and garlic fine. Heat olive oil in a soup pot and sauté onions and garlic until golden. Add basil, oregano and tomato paste. Cook one minute longer. Stir well.

Put aside hairy sprigs of fennel and save for salad. Cut the fennel stalks and bulb into chunks. Chop in a food processor or blender. Add tomato and potato, continuing to work the machine. Add vegetable cooking water, a little at a time, pureeing until a soup-like consistency is achieved.

Transfer soup to pot. Warm slowly but do not allow the soup to come to a simmer. Stir frequently. Season to taste with salt and pepper.

Garnish: Cooked macaroni and fresh, grated Parmesan cheese, or use soy Parmesan.

Fennel Soup Spicy Style

1/2 small onion
2 cloves garlic
1" piece ginger
1 Tbsp. mustard seed
1 Tbsp. cumin seed
1-1/2 Tbsp. fennel seed
Salt and pepper to taste
2 large fennel bulbs
3 stalks celery
1 small carrot
1 small steamed potato
4 cups vegetable bouillon

Heat oil of choice in a soup pot. Chop onion fine and sauté until golden, stirring often. In a dry frying pan, roast mustard, cumin and fennel seeds for three minutes over high flame. Grind the roasted seeds in a food mill and add to the sautéd onion. Peel ginger. Place in blender with 1 tablespoon water and blend. Add to pot along with ground seeds. Cook for one minute longer. Stir well.

Remove stems and sprigs from fennel bulbs. Save sprigs for the garnish. Cut fennel bulbs, celery stalks, carrot and potato into 1" pieces.

Chop fennel bulb in a food processor or blender. Add celery, carrot and potato pieces a few at a time and continue chopping. Use a little of the bouillon to keep the blade moving and chop until the ingredients are assimilated. Add the remainder of the bouillon slowly, continuing to process the machine until a soup-like consistency is achieved.

Transfer the soup to the pot. Warm slowly but do not allow the soup to reach a simmer. Stir often. Salt and pepper to taste.

Garnish: Sprigs of fennel

Green Bean Almond Soup

1 cup almonds
4 cups water
1/2 small onion
1 Tbsp. margarine or butter
1/4 tsp. coriander
1/4 tsp. curry powder
1 lb. green beans
1 small carrot
1 small steamed potato
Salt to taste
Pinch of cayenne pepper

Soak almonds in water to cover overnight. Discard the soaking water. Chop almonds in a food processor or blender. Add fresh water a little at a time, continuing to chop. If you prefer a thick-textured soup, use this as your soup base. If a smoother texture is desired, strain the almond mixture and remove the pulp. Use the almond liquid as your soup base.

Heat margarine or butter in a soup pot. Chop onion fine and sauté until transparent. Add coriander and curry. Stir well.

Trim green beans and cut in half. Chop in a food processor or blender. Cut the carrot into 1" pieces. Add carrot and potato to the work bowl, processing until the ingredients are assimilated. Add almond soup base a little at a time, pureeing until a soup-like consistency is achieved.

Transfer soup to pot. Warm slowly over low heat, stirring until serving temperature but do not let the soup reach the simmer point. Season to taste with salt and pinch of cayenne pepper.

Garnish: Chopped cilantro

Green Bean Soup Anise Flavor

1 Tbsp. margarine or butter
1 small onion
1 Tbsp. anise seed
1 lb. green beans
1 medium steamed potato
4 cups chicken-flavored bouillon
Salt and pepper

Heat margarine or butter in a soup pot. Chop onion fine and sauté until transparent. Stir often.

In a dry frying pan, roast anise seeds for a few minutes. Grind roasted seeds in a food mill and add to the sauté. Cook one minute longer. Stir well.

Trim green beans and cut in half. Chop in a food processor or blender. Add potato and process until the ingredients are merged. Use a little of the bouillon to keep the blade moving. Add the remaining bouillon slowly, pureeing until a soup-like consistency is achieved.

Remove soup to pot. Warm just below simmer point. Stir frequently. Season to taste with salt and pepper.

Garnish: Cream cheese or soy cream cheese

Green Bean Basil Soup

1 Tbsp. oil of choice (olive, canola, margarine, butter)
1 small onion
1 lb. fresh green beans
1/2 cup fresh basil leaves
1 medium steamed potato
4 cups leftover vegetable cooking water or fresh water
2 Tbsp. lemon juice
Salt and pepper to taste

Heat oil in soup pot. Sauté onions in oil, stirring frequently until onions are slightly browned.

Tip green beans and cut them in half. Chop green beans in a food processor or blender a few at a time. Add basil leaves and continue chopping. Add steamed potato and continue to chop. Add water a little at a time, processing until a pureed consistency is achieved. Pour puree into the soup pot. Add lemon juice.

Warm with low temperature to just before soup begins to simmer, stirring frequently. Season with salt and pepper to taste.

Garnish: Thinly-sliced scallions

Green Bean Soup Simple

1/2 lb. green beans
1 medium steamed potato
4 cups mock cream of chicken soup (see Introduction)
1/4 tsp. nutmeg
Salt and pepper to taste

Snip tips from green beans and discard. Cut beans into 2″ lengths and chop in a food processor or blender. Quarter the potato and combine it with the beans, continuing to chop until the ingredients are assimilated. Use a little of the mock cream-of-chicken soup to keep the blade moving. Add the remainder of the soup and continue to process until thoroughly pureed.

Transfer soup to a kettle and warm slowly. Do not allow soup to reach a simmer. Stir often. Season to taste with salt and pepper.

Garnish: Grated hard-boiled egg

Green Bean Soup Cream Style

4 cups milk product of choice (soy milk evaporated
   skim milk, milk, cream)
1/8 cup sherry
1/8 tsp. thyme
1/8 basil
1/8 tsp. marjoram
1 tsp. sugar
1 tsp. celery salt
1 lb. green beans
1 small steamed potato
Salt and pepper to taste

Warm milk product of choice. Add sherry, thyme, basil, marjoram, sugar and celery salt. Cook one minute longer. Stir well.

Snip tips off green beans and discard. Cut green beans into 1" pieces and chop in a food processor or blender. Add potato and process until the ingredients are assimilated. Use a little of the warmed milk product of choice to keep the blade moving. Add the remaining milk product of choice slowly, pureeing until a soup-like consistency is achieved.

Transfer soup to a kettle. Warm over low heat to serving temperature. Do not allow the soup to simmer. Season to taste with salt and pepper.

Garnish: Toasted almonds, thinly sliced

Green Bean Soup Tomato Flavor

1 small onion
1 Tbsp. oil of choice (olive, canola, margarine, butter)
1/2 lb. green beans
5 fresh basil leaves
1 medium steamed potato
1 small, very ripe tomato
4 cups tomato-flavored stock (see Introduction for
    directions)
Salt and pepper to taste

Chop onion fine. Heat oil of choice in a kettle. Add onion and sauté until the onion is lightly browned. Stir frequently.

Remove the tips of the green beans and cut into 1" pieces. Chop in a food processor or blender. Add the basil leaves and steamed potato, continuing to work the machine until the ingredients have been assimilated. Add the tomato-flavored stock, a little at a time, and puree until a soup-like consistency is achieved.

Transfer soup to the kettle. Warm just below the simmer point. Stir frequently. Season to taste with salt and pepper.

Garnish: Shredded arugula leaves

Green Pea Asparagus Soup

2 Tbsp. oil of choice (olive, canola, margarine, butter)
4 Tbsp. oatmeal
1 Tbsp. Spike (seasoning available in health food stores)
1 lb. asparagus spears
1 cup shelled green peas
4 cups vegetable bouillon

Heat oil of choice in a soup pot. Add oatmeal and Spike. Cook slowly until the ingredients are assimilated. Stir well.

Remove tough ends from the asparagus and save for stir-frying. Chop asparagus spears in a food processor or blender. Add green peas and the vegetable bouillon slowly. Work the machine until a soup-like consistency is achieved.

Remove soup to pot. Over low heat, warm thoroughly. Do not allow the soup to reach a simmer. Stir frequently. Adjust seasoning with additional Spike if desired.

Garnish: Whole, raw green peas

Green Pea Soup Cream Style

1/2 small onion
1 Tbsp. oil of choice (olive, canola, margarine, butter)
1/4 tsp. each dried basil, chives,
    thyme, parsley, dill
2 cups milk product of choice (soy milk, evaporated
    skim milk, milk, cream)
1 1/2 cups shelled green peas
1 medium steamed potato
2 cups chicken-flavored bouillon
Salt and pepper

Chop onion fine. Heat oil of choice in a kettle and sauté the onion until transparent. Add dried herbs and milk product of choice. Cook a few minutes longer on a low flame. Stir well. Allow to cool.

Chop the shelled green peas in a food processor or blender. Quarter the potato and add it to the work bowl, continuing to chop until the ingredients are assimilated. Use small amounts of the bouillon to keep the blade moving. Add the remaining bouillon, processing until a soup-like consistency is achieved.

Transfer soup to kettle and warm slowly. Do not allow the soup to reach a simmer. Taste and adjust the seasoning with salt and pepper.

Garnish: **Mustard Onion Croutons**

4 slices of day-old, whole wheat bread.
3 tablespoons oil of choice
1-1/2 tablespoons finely grated onion
1-1/2 tablespoons prepared mustard

Heat oil of choice in a small pan, add onion and mustard. Cook until onion is tender and the ingredients assimilated. Stir well. Spread mixture on bread and place in a preheated 350º oven for 10 minutes. Cut into crouton-size pieces.

Green Pea Soup
Ginger Flavor

1/2 small onion
1 clove garlic
2″ piece ginger
1 Tbsp. oil of choice (olive, canola, margarine, butter)
    or plain water
1 tsp. tarragon
1/4 tsp. ground cloves
1-1/2 cup shelled green peas
1 medium steamed potato
4 cups leftover vegetable cooking water
Salt and pepper to taste

Chop onion and garlic fine. Heat oil of choice in a kettle. Sauté onion and garlic until soft. Peel ginger and chop in a blender with 1 Tbsp. water. Add ginger, tarragon and cloves to the sauté and cook one minute longer. Stir well.

Chop green peas in a food processor or blender. Add potato and process. Add leftover vegetable cooking water, a little at a time, working the machine until a soup-like consistency is achieved.

Transfer soup to kettle. Warm just below simmer point. Stir frequently. Season to taste with salt and pepper.

Garnish: Chopped mint leaves

Green Pea Soup Oriental Style

2 Tbsp. toasted sesame oil
1/2 small onion
2 cloves garlic
1/2" piece fresh ginger
1-1/2 cups shelled green peas
1 small steamed potato
4 cups chicken-flavored bouillon
1 Tbsp. tamari

Heat sesame oil in a kettle. Chop garlic and onion fine. Add them to the kettle and cook until the onion is soft. Stir frequently. Peel ginger. Place in blender with one-half tablespoon water and blend. Add blended ginger to kettle.

Chop green peas and boiled potato in a food processor or blender until they are assimilated. Add the bouillon and tamari, a little at a time, continuing to work the machine until a soup-like consistency is achieved.

Remove soup to kettle. Warm slowly, but do not allow the soup to come to a simmer. Stir often. Adjust the seasoning to your taste with additional tamari.

Garnish: Cooked Udon noodles (Japanese noodles)

Green Pea Sesame Seed Soup

1/2 small onion
1 stalk celery
1/4 tsp. celery seed
1 Tbsp. toasted sesame oil
1-1/2 cups shelled green peas
1/2 cup sesame seeds
4 cups chicken-flavored bouillon
Tamari to taste

Chop onion and celery fine. Add toasted sesame oil along with celery seed and heat in a soup kettle until the onion and celery are soft.

Chop green peas in a food processor or blender. Pulverize sesame seeds in a food mill and add to the green peas. Continue to process the machine until the ingredients are assimilated. Add the bouillon, a little at a time, and puree until a soup-like consistency is achieved.

Transfer soup to kettle and warm slowly. Do not allow the soup to come to a simmer. Stir frequently. Add tamari to taste.

Garnish: Sesame sticks

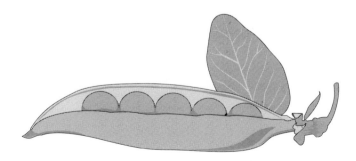

Green Pea Soup Simple

1 medium onion
1 Tbsp. oil of choice (olive, canola, margarine, butter)
1/2 cup white wine
Pinch of sugar
1 small carrot
1 cup shelled green peas
6 mint leaves
1 medium steamed potato
4 cups chicken-flavored bouillon
Salt and pepper to taste

Chop onion fine. Heat oil of choice in a kettle and sauté onion until transparent. Add white wine and sugar. Stir well.

Cut carrots into 1" chunks and chop in a food processor or blender. Add green peas, steamed potato and mint leaves a few at a time. Continue chopping until the ingredients are assimilated. Use a little of the bouillon to keep the blade moving. Add the remaining bouillon processing until a soup-like consistency is achieved.

Remove soup to kettle and warm just below the simmer point. Stir often. Season to taste with salt and pepper.

Garnish: Thinly sliced red radishes

Green Pea Soup Spicy Style

4 cups chicken-flavored bouillon
1/2 tsp. grated ginger
1 tsp. coriander
1 tsp. cumin
1 tsp. cardamom
1/8 tsp. cloves
1/4 tsp. cinnamon
1 tsp. turmeric
Dash of cayenne
2 cups shelled green peas
2 tomatoes
1 medium steamed sweet potato

Combine spices with bouillon and heat. Stir often. Allow to cool.

Chop tomatoes and green peas in a blender or food processor. Quarter sweet potato and add to the mixture. If necessary use a little of the cooled bouillon to keep the blade moving. Process the ingredients until they are assimilated. Add the remaining bouillon, a little at a time, and continue to work the machine until a soup-like consistency is achieved.

Garnish: Additional shelled raw green peas

Green Pea Tomato Soup

1 Tbsp. oil of choice (olive, canola, margarine, butter)
3 cloves garlic
1 small onion
2 Tbsp. marjoram
2 bay leaves
1/2 cup white wine
1-1/2 cups shelled green peas
1 small steamed potato
4 medium tomatoes
1 small carrot
2 stalks celery
4 cups chicken-flavored bouillon
Salt and pepper to taste

Heat oil of choice in a kettle. Chop onion and garlic fine. Sauté until soft. Crush bay leaves in a food mill. Add bay leaves, marjoram and white wine. Cook one minute longer. Stir well.

Chop shelled green peas in a food processor or blender. Quarter tomatoes and potato and add them to the work bowl, continuing to process. Use small amounts of the bouillon to keep the blade moving. Discard the stems of the parsley (or munch on them while working!). Add the sprigs to the other ingredients and chop until assimilated. Add the remaining bouillon, and puree until a soup-like consistency is achieved.

Transfer soup to kettle. Warm slowly. Do not allow the soup to reach a simmer. Stir often. Season to taste with salt and pepper.

Garnish: Gamasio (sea salt and toasted sesame seeds ground together—available in health food stores)

Kale Basil Soup

1/2 medium onion
3 cloves garlic
1 Tbsp. oil of choice (olive, canola, margarine, butter)
1 cup basil leaves
4 cups baby kale leaves
1 medium steamed potato
4 cups chicken-flavored bouillon

Mince onion and garlic. Heat oil of choice in a kettle.
Sauté onions and garlic until tender.

Chop basil leaves in a food processor or blender. Add
the baby kale leaves and steamed potato and continue
chopping. Add the bouillon, a little at a time, and puree
until the mixture reaches a soup-like consistency.

Pour soup into kettle and warm slowly, until just below
the simmer point. Stir often. Season to taste with salt and
pepper.

Garnish: 1/4 cup toasted pine nuts, freshly grated
Parmesan cheese or soy Parmesan

Kale Soup Indian Style

First create this soup base from red lentils, also called dahl:

1 Tbsp. canola oil
1/2" piece of ginger
1/2 small onion*
2 cloves garlic*
1/4 tsp. coriander
1/4 tsp. chili powder
Dash of cayenne
Salt to taste
4 cups red lentils (dahl)
5 cups water
1/4 tsp. cumin

Can substitute 1/4 tsp. asafoetida (see Introduction)

1 pound fresh kale

Heat oil in a kettle. Chop ginger, onion and garlic fine. Sauté until soft. Combine with the remaining spices and water. Bring to a boil over moderate heat. While water is heating, sort through the dahl and remove any foreign matter. Rinse the dahl to remove dust and transfer to the kettle. Cook over moderate heat for 30 minutes with the lid on. Season to taste with salt. Allow to cool and this will be your soup base.

Thoroughly wash kale. Remove the stems and discard them. If you use baby kale, you can use the stems. Chop the kale in a food processor or blender, while adding small amounts of the cooled dahl to keep the blade moving. Add remaining dahl and continue to work the machine until a soup-like consistency is achieved.

Transfer the soup to a kettle and warm slowly. Do not allow the soup to reach a simmer. Stir frequently. Taste and adjust the seasoning with salt.

Garnish: Fresh chopped cilantro and lemon wedges. If served with rice, this soup is a complete meal.

Kale Soup Portuguese Style

2 cups white beans
5 cups chicken-flavored bouillon
1 lb. kale (use baby kale, if possible)
1/8 tsp. saffron
Salt and pepper to taste

In this recipe, the cooked white beans and liquid become the soup base.

Soak beans overnight in water to cover. Discard soaking water. Add 5 cups chicken-flavored bouillon to beans. Cook until very tender, approximately 2 hours. Puree half of the cooked beans and all of the liquid in a food processor or blender. This will be your soup base. Reserve the whole beans.

If using regular kale, remove thick stems and tear into 1" pieces. If using baby kale, this step is not necessary. Chop the kale in a food processor or blender. Add the bean soup base, a little at a time, continuing to process the machine until a soup-like consistency is achieved.

Transfer the soup to a kettle. Rub the saffron between your finger and mix it with a tablespoon of water. Combine this mixture and the reserved whole beans with the soup. Warm slowly to just below the simmer point. Stir frequently. Season to taste with salt and pepper.

Garnish: Cooked macaroni

Kohlrabi Almond Soup

1 cup almonds
4 cups water
1/2 small onion
1 Tbsp. margarine or butter
1 tsp. cardamum seeds
3/4 lb. kohlrabi
1 small steamed potato
Salt and pepper to taste

Soak almonds overnight in water to cover. Discard the soaking water. Chop almonds in a food processor or blender. Add fresh water slowly, continuing to process. Strain the almond mixture and remove the pulp. Save for dessert topping. Use the almond liquid as your soup base.

Heat margarine or butter in a soup pot. Chop onion fine and sauté until transparent. Toast cardamom seeds in a dry skillet. Grind in a food mill. Add to sauté. Cook one minute longer. Stir well.

Peel kohlrabi and cut into 1" chunks. Chop in a food processor or blender. Add steamed potato, chopping until ingredients are blended. Add almond liquid slowly processing until a soup-like consistency is achieved.

Transfer soup to pot. Warm over low heat, stirring until serving temperature. Do not let the soup reach the simmer point. Season to taste with salt and pepper.

Garnish: Shelled raw green peas

Kohlrabi Soup Cream Style

1 small onion
1 Tbsp. oil of choice (olive, canola, margarine, butter)
1/4 tsp. nutmeg
3/4 lb. kohlrabi
1 large steamed potato
3 cups chicken-flavored bouillon
1 cup milk product of choice (soy milk, evaporated
     skim milk, milk, cream)
Salt and pepper to taste

Heat oil of choice in a soup pot. Chop onion fine and
sauté until transparent. Add nutmeg and milk product of
choice. Stir well.

Peel kohlrabi and cut into 1" chunks. Chop in a food
processor or blender. Add potato and process. Add
bouillon, working the machine until a soup-like
consistency is achieved.

Remove soup to pot. Warm over low heat, stirring until
serving temperature, but do not let the soup reach the
simmer point. Season to taste with salt and pepper.

Garnish: **Whole wheat bread croutons with Parmesan
cheese or soy Parmesan**

Spread day-old whole wheat bread with oil of choice.
Sprinkle with Parmesan cheese. Place in broiler until
cheese melts. Cut into bite-sized pieces.

Kohlrabi Soup Zesty Style

1 Tbsp. olive oil
1 medium onion
4 cloves garlic
1 tsp. basil
1/4 tsp. chili powder
1/2 tsp. curry
Salt to taste
1/4 tsp. grated lemon rind
2 cups coconut milk
1 lb. kohlrabi
1 small steamed potato
2 cups leftover vegetable cooking water or plain water

Heat olive oil in a soup pot. Chop onion and garlic fine. Sauté until lightly brown. Add basil, chili powder, curry, grated lemon rind and coconut milk. Cook one minute longer. Stir well.

Peel kohlrabi and discard leaves. Cut into 1" chunks. Chop in a food processor or blender. Add potato and process until the ingredients are assimilated.

Add the leftover vegetable cooking water, a little at a time, chopping until a soup-like consistency is achieved.

Transfer soup to pot. Warm slowly over low heat to serving temperature. Do not allow the soup to reach a simmer. Stir often. Adjust seasoning with salt.

Garnish: Fresh corn kernels

Kohlrabi Tomato Soup

1 Tbsp. oil of choice (olive, canola, margarine, butter)
1/2 medium onion
1 tsp. thyme
1/2 tsp. cumin
2 Tbsp. tomato paste
1 lb. kohlrabi
1 medium, very ripe tomato
1 small steamed potato
4 cups chicken-flavored bouillon
Salt and pepper to taste

Heat oil of choice in a kettle. Chop onion fine and sauté until translucent. Add thyme, cumin and tomato paste. Cook one minute longer. Stir well.

Peel kohlrabi and discard leaves. Cut into 1'' chunks. Chop in a food processor or blender. Quarter tomato and potato. Add to the work bowl, processing until the ingredients are assimilated. Use a little of the bouillon to keep the blade moving. Add the remaining bouillon slowly, continuing to chop until a soup-like consistency is achieved.

Remove soup to kettle. Warm just below simmer point. Stir frequently. Season to taste with salt and pepper.

Garnish: Cashew nuts

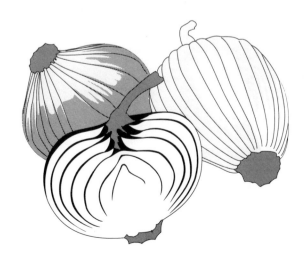

Lettuce Celery Soup

2 large leeks
1-1/2 Tbsp. margarine or butter
1 Tbsp. lemon juice
3 celery stalks
1 head butter lettuce
1 medium steamed potato
4 cups chicken-flavored bouillon
Salt and pepper to taste

Slice white part of leek thin and sauté in margarine or butter until soft. Stir often. Add lemon juice. Cook one minute longer.

Remove the core and any tough stem portions from the lettuce. Wash thoroughly to remove grit. Chop the lettuce leaves in a food processor or blender. Add potato and process until the ingredients are assimilated. Add bouillon, a little at a time, continuing to chop until a soup-like consistency is achieved.

Transfer the soup to a kettle. Add sauté. Warm just below the simmer point. Stir frequently. Adjust seasoning with salt and pepper.

Garnish: Thinly sliced red radishes

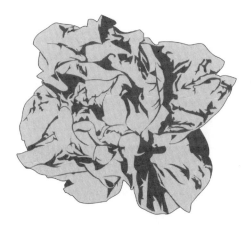

Lettuce Green Pea Soup

1 small onion
1 Tbsp. oil of choice (olive, canola, margarine, butter)
3/4 tsp. dried thyme
1 cup shelled green peas
1 large bunch romaine, or other dark green lettuce
1 medium steamed potato
4 cups mock-chicken soup (see Introduction)
1/3 cup white wine
Salt and pepper to taste

Chop onion fine. Heat oil of choice in a soup kettle. Add onion, dried thyme and white wine. Cook until the onion is soft. Stir often.

Remove core from lettuce and discard. Thoroughly clean grit and dirt from lettuce leaves. Chop lettuce in a blender or food processor. Quarter the potato. Add the potato and peas to the bowl and continue to work the machine until the ingredients are thoroughly assimilated. You may need a little of the mock chicken soup to keep the blades moving. Add the remaining soup, a little at a time, and puree until a soup-like consistency is achieved.

Transfer soup to kettle and warm slowly. Do not allow soup to reach a simmer. Stir frequently. Season to taste with salt and pepper.

Garnish: A dollop of yogurt or soy yogurt

Lettuce Soup Cream Style

1/2 small onion
1 Tbsp. oil of choice (olive, canola, margarine, butter)
1 small head romaine lettuce
4 arugula leaves
1 medium steamed potato
3 cups chicken-flavored bouillon
1 cup buttermilk (or soy milk with 3/4 Tbsp. lemon
    juice)
1 tsp. sugar
Salt and pepper to taste

Chop onion fine and sauté in oil of choice until translucent. Separate lettuce leaves and wash thoroughly. Tear into 2″ pieces Follow the same procedure with the arugula leaves. Chop lettuce and arugula leaves in a food processor or blender. Add the potato and continue chopping until the ingredients are assimilated. Add the bouillon and buttermilk, a little at a time, processing the machine until a soup-like consistency is achieved.

Remove the soup to a kettle. Add the sauté and sugar. Warm over low heat.

Stir frequently. Do not let soup reach the simmer point. Season to taste with salt and pepper.

Garnish: Slivered toasted almonds

Lettuce Soup Simple

1 small onion
2 cloves garlic
1 Tbsp. oil of choice (olive, canola, margarine, butter)
1 tsp. tarragon
1 head romaine lettuce
1 medium steamed potato
4 cups vegetarian white stock (see Introduction)
Salt and pepper to taste

Chop onion and garlic fine. Heat oil of choice in a soup pot. Add onion and garlic; sauté until lightly browned. Add tarragon. Stir well.

Wash lettuce leaves and tear them into 2" pieces. Chop in a food processor or blender. Add the potato and continue chopping until the ingredients are assimilated. Add the vegetarian white stock, a little at a time, pureeing until the soup-like consistency is achieved.

Remove soup to pot. Warm over low heat to just below simmer point. Season to taste wit salt and pepper.

Garnish: **Bulgur**

Soak 1/2 cup bulgur (specially processed cracked wheat) in cold water for 30 minutes. Drain and add to soup.

Mushroom Kasha Soup

1/2 pound fresh mushrooms (Shiitake mushrooms
    taste great)
4 Tbsp. onion soup mix (soup base)
1 cup kasha (roasted bulgur)
4 cups water or leftover vegetable cooking water
1 medium steamed potato
2 cups water
Tamari to taste

Clean mushrooms. An efficient, quick cleaning method: put one tablespoon white vinegar in a plastic bag along with the mushrooms; add warm water and agitate. Using a sharp instrument, poke holes in the plastic bag and allow the vinegar-water to run out. *Voila!* Clean mushrooms! Remove the stems from the mushrooms and save for other recipes, e.g., stews, scrambled eggs and mushrooms, etc.

Add onion soup mix to water to create the soup base. Puree the mushrooms, a few at a time, in the food processor or blender. Add a little of the soup base and continue working the machine. Add the steamed potato and continue pureeing. Keep adding small amounts of the soup base to the puree and continue processing until a soup-like consistency is achieved.

Pour the soup into kettle. Warm to serving temperature but do not allow soup to simmer. Stir often. Add tamari to taste.

Kasha Recipe

Boil two cups of water. Add the roasted bulgur, cover and simmer on a low heat for 15 minutes. Add the kasha to the soup.

Garnish: Raw green peas

Mushroom Soup Chinese Style

1" piece fresh ginger
1 medium onion
4 cloves garlic
1/2 tsp. chili powder
1-1/2 Tbsp. toasted sesame oil
8 oz. mushrooms (Shiitakes are delicious)
1 medium steamed potato
4 cups chicken-flavored bouillon
1/4 cup rice vinegar
3 Tbsp. tamari or to taste

Chop ginger, onion and garlic fine. Heat sesame oil in a kettle. Add ginger, onion, garlic and chili powder. Cook until ginger, onion and garlic are soft. Stir frequently.

Remove stems from mushrooms. Do not use but save for stir-frying. Chop mushrooms in a food processor or blender. Add steamed potato and continue chopping. Add the bouillon, a little at a time, and puree the ingredients until a soup-like consistency is achieved. Add rice vinegar and tamari; whirl the machine once more.

Transfer soup to kettle. Warm just below the simmer point. Stir frequently. Adjust seasoning to taste.

Garnish: Fresh bean sprouts

Mushroom Orange Soup

1 Tbsp. oil of choice (olive, canola, margarine, butter)
1/2 small onion
1 clove garlic
1 tsp. cumin
1 tsp. coriander
1/2 tsp. cinnamon
1 tsp. mustard
1-1/2" piece ginger
3/4 lb. white button mushrooms
1 medium steamed potato
3 cups vegetable bouillon
1 cup fresh orange juice
1/4 cup sherry
Salt and pepper to taste

Heat oil of choice in kettle. Chop onion and garlic fine and sauté until soft. Add cumin, coriander, cinnamon and mustard. Cook one minute longer. Peel ginger and place in a blender with a little water to cover, and chop ginger fine. Add this mixture and sherry to kettle and cook one minute longer. Stir well.

Clean mushrooms. Remove and discard stems. An efficient, quick cleaning method: put one tablespoon white vinegar in a plastic bag along with the mushrooms; add warm water and agitate. Using a sharp instrument, poke holes in the plastic bag and allow the vinegar-water to run out. *Voila!* Clean mushrooms!

Chop mushrooms in a food processor or blender. Add potato and continue to chop. Use a little of the bouillon to keep the blender moving. Add orange juice and process. Add remaining bouillon, pureeing until a soup-like texture is achieved.

Remove soup to kettle. Warm just below simmer. Stir frequently. Season to taste with salt and pepper.

Garnish: Finely grated orange rind

Mushroom Tomato Soup

1 Tbsp. oil of choice (olive, canola, margarine, butter)
1 medium onion
2 cloves garlic
3 Tbsp. Picante sauce (store bought)
1 lb. white button mushrooms
1 medium steamed potato
3 medium, very ripe tomatoes
8 sprigs parsley
4 cups vegetable bouillon
Salt and pepper

Heat oil of choice in kettle. Chop onion and garlic fine.
Sauté until slightly brown. Add Picante sauce. Stir well.
Cook one minute longer.

Clean mushrooms (see previous page for cleaning tip).
Discard stems. Chop mushroom caps in food processor or
blender. Cut tomatoes and potato into small chunks. Add
to work bowl and process. Discard parsley stems. Chop
parsley leaves with other ingredients until assimilated.
Use a little of the bouillon to keep blade moving. Add
remaining bouillon slowly, pureeing until a soup-like
consistency is achieved.

Remove soup to kettle. Heat just below simmer point.
Stir frequently. Adjust seasoning with salt, pepper and
additional Picante sauce.

Garnish: Garlic croutons

Mushroom Soup Russian Style

2 leeks
1-1/2 Tbsp. margarine or butter
2 bay leaves
1 cup milk product of choice (soy milk, evaporated
    skim milk, milk, cream)
1 lb. white button mushrooms
1 bunch dill
2 medium carrots
1 large steamed potato
3-1/2 cups vegetable bouillon
Salt and pepper to taste

Use white part of leeks and slice thin. Heat margarine
or butter in a soup pot and sauté leeks until they are
transparent. Crush bay leaves in a food mill. Combine
with sauté. Add milk product of choice and cook a few
minutes longer until milk is warm. Stir well.

Remove stems from mushrooms and save for stir-frying.
Clean caps and chop in a food processor or blender. Cut
carrots into 1" pieces and add to the work bowl; process.
Discard stems from one bunch dill. Combine dill leaves
and potato with other ingredients, chopping until
assimilated. Use a little of the bouillon to keep the blade
moving. Add remaining bouillon, pureeing until a
soup-like consistency is achieved.

Transfer soup to pot. Warm slowly over low heat to
serving temperature. Stir frequently. Do not allow soup to
reach a simmer point. Season to taste with salt and
pepper.

Garnish: Herbed croutons

Mushroom Soup Simple

1 small onion
2 cloves garlic
1 Tbsp. oil of choice (olive, canola, margarine, butter)
1 tsp. thyme
1 tsp. table mustard
1/2 cup white wine
1 lb. white button mushrooms
1 small steamed potato
1 cup milk product of choice (soy milk, evaporated
    skim milk, milk, cream)
3 cups chicken-flavored bouillon
Salt and pepper to taste

Chop onion and garlic fine. Heat oil of choice in soup pot and sauté until soft. Add thyme, table mustard and wine. Cook for one minute longer. Stir well.

Remove stems from mushrooms and discard. Clean mushrooms (see suggestion for cleaning in Mushroom Kasha Soup recipe). Chop mushrooms in a food processor or blender. Add potato and process. Use a little of bouillon to keep blade moving. Add remaining bouillon and milk product of choice, pureeing until a soup-like consistency is achieved.

Transfer soup to pot. Warm over low heat, stirring until serving temperature but do not let it simmer. Season to taste with salt and pepper.

Garnish: Cooked lima beans

Mushroom Sesame Soup

1 medium onion
4 cloves garlic
1-1/2 Tbsp. toasted sesame oil
2 tsp. powdered ginger
1 lb. white button mushrooms
1 medium tomato
1 stalk celery
2/3 cup tahini (sesame butter)
4 cups leftover vegetable cooking water or plain water
Tamari to taste

Chop onion and garlic fine. Heat sesame oil in a kettle. Sauté onion and garlic until lightly browned. Add ginger and cook one minute longer. Stir well.

Remove stems from mushrooms and save for stir-frying. Clean caps (see Mushroom Kasha recipe for tip) and chop in a food processor or blender. Quarter tomato and cut celery into 1'' chunks. Add them to the work bowl and process. Add tahini, continuing to chop until ingredients are combined. Use a little of the vegetable cooking water to keep the blade moving. Add remaining liquid, pureeing until a soup-like texture is achieved.

Transfer soup to kettle. Warm slowly over low heat to serving temperature, stirring frequently; do not allow to simmer. Season to taste with tamari.

Garnish: Sesame sticks

Parsley Soup Simple

1 bunch parsley
1 large potato steamed
4 cups garlic–flavored stock (see Introduction)
1 Tbsp. Spike (seasoning available in health stores)

Remove stems from parsley and discard. Cut potato into 1″ pieces. Chop parsley in a food processor or blender. Add potato, continuing to work machine until ingredients are assimilated. Use a little of the garlic-flavored stock to keep blade moving. Add remaining stock, a little at a time, pureeing until a soup-like consistency is achieved.

Remove soup to a kettle. Warm slowly. Stir often. Do not allow soup to reach a simmer. Add Spike and adjust seasoning to taste.

Garnish: Raw carrot curls

Parsley Soup Swiss Style

1 large onion
1-1/2 Tbsp. oil of choice (olive, canola, margarine,
    butter)
1 Tbsp. table mustard
1/8 tsp. white pepper
2 medium potatoes
3 cups leftover vegetable cooking water, or plain water
1 cup milk product of choice (soy milk, evaporated
    skim milk, milk, cream)
Salt to taste

Chop onion fine. Heat oil of choice in a kettle. Sauté onion until transparent. Add mustard and pepper. Stir well.

Chop parsley leaves and stems in a food processor or blender. Cut potatoes into 1" chunks and add to work bowl, a few at a time, continuing to chop. Add a little of the vegetable cooking water to keep blade moving. Adding remaining vegetable cooking water and milk product of choice, processing until a soup-like consistency is achieved.

Remove soup to kettle. Warm over low heat, stirring often. Bring soup to just below simmer point. Season to taste with salt.

Garnish: Grated Swiss cheese

Parsley Carrot Soup

1/2 small onion
1 Tbsp. oil of choice (olive, canola, margarine, butter)
2 medium carrots
1 bunch parsley
1 small steamed potato
4 cups chicken-flavored bouillon
1 Tbsp. miso

Mince onion. Heat oil of choice in a kettle, add onion and brown.

Cut carrots into 2" chunks and chop finely in a food processor or blender. Discard stems from parsley and add leaves to the work bowl; continue chopping. Add steamed potato and continue to work machine. Slowly add chicken-flavored bouillon, pureeing until a soup-like consistency is achieved. Add miso and whirl machine around once more.

Pour soup into kettle. Warm and stir frequently, but do not allow soup to come to a simmer.

Garnish: Cooked vermicelli

Parsnip Apricot Soup

1 small onion
2 cloves of garlic
1 Tbsp. oil of choice (olive, canola, margarine, butter)
3 medium parsnips
1 small steamed potato
2 cups orange juice
2 cups vegetable bouillon
1 cup milk product of choice (soy milk, evaporated
    skim milk, milk, cream)
1 cup dried apricots reconstituted*
Salt and pepper to taste

To reconstitute apricots, cover them with water and soak overnight.

Chop onion and garlic fine. Heat oil of choice in a kettle. Sauté onion and garlic until soft. Stir often.

Peel, trim and cut parsnips into 1" pieces. Chop parsnips in food processor or blender. Add potato, working the machine. Use a small amount of bouillon to keep blade moving.

Add apricots and soaking water to work bowl, processing until ingredients are well blended. Add remaining bouillon, orange juice and milk product of choice slowly, continuing to chop until a soup-like consistency is achieved.

Transfer soup to the kettle. Stir well. Warm just below simmer point. Season to taste with salt and pepper.

Garnish: Freshly grated nutmeg

Parsnip Soup Curry Flavored

1 Tbsp. oil of choice (butter, margarine, olive, canola)
1/4 cup oatmeal
1 Tsp. curry powder
2 medium parsnips
4 cups vegetable bouillon
1" piece ginger
Salt to taste

Heat oil of choice in soup pot. Add oatmeal and curry powder. Stir until curry is absorbed and oatmeal turns amber color.

Peel ginger. Place in blender with one tablespoon water and blend. Add blended ginger to soup pot.

Peel, trim and cut parsnips into 1" pieces. Chop parsnips in food processor or blender. Add bouillon, a little at a time, continuing to work machine until a soup-like consistency is achieved. Add this mixture to soup pot.

Warm, stirring frequently. Do not allow soup to simmer. Salt to taste.

Garnish: Chopped parsnip leaves. If not available use parsley.

Parsnip Soup Fennel Flavor

1/2 medium onion
1 Tbsp. oil of choice (olive, canola, margarine, butter)
1 tsp. fennel seeds
3 parsnips
1 small steamed potato
4 cups vegetable bouillon
Salt and pepper to taste

Chop onion fine. Heat oil of choice in a soup kettle. Sauté onion until translucent. In a dry skillet, toast fennel seeds for a few moments over high flame. Do not let them burn. Chop toasted fennel seeds in a food mill until pulverized. Add to sauté and stir well.

Cut parsnips into 1" pieces. Chop in food processor or blender. Add potato and process until the ingredients are assimilated. Use a little of the bouillon to keep blade moving. Add remaining bouillon slowly, chopping until a soup-like consistency is achieved.

Remove soup to kettle. Warm slowly to serving temperature, stirring frequently. Do not allow soup to simmer.

Garnish: Fresh fennel leaves

Parsnip Dill Soup

1 medium onion
3 cloves garlic
1 Tbsp. oil of choice (olive, canola, margarine, butter)
1 Tbsp. dry dill
3 medium parsnips
2 stalks celery
1 small steamed potato
4 cups mock cream-of-chicken soup base
    (see Introduction)
Salt and pepper to taste

Chop onion and garlic fine. Heat oil of choice in a soup pot. Add onion, garlic and dill. Sauté until the onion and garlic are soft. Stir often.

Peel, trim and cut parsnips into 1" pieces. Cut celery into 1" pieces. Chop parsnips and celery in a food processor or blender. Add potato and process the machine until ingredients are well blended. Use small amounts of mock cream-of-chicken soup base to keep blade moving. Add remaining soup base, a little at a time, continuing to chop until a soup-like consistency is achieved.

Transfer soup to pot. Warm slowly, stirring often, just to below simmer point. Season to taste with salt and pepper.

Garnish: Sprigs of fresh dill

Parsnip Soup Cream Style

1 medium onion
1 Tbsp. oil of choice
1" piece of ginger
3 parsnips
1 small steamed potato
3 cups vegetable bouillon
1 cup milk product of choice (soy milk, evaporated
    skim milk, milk, cream)
Tamari to taste

Chop onion fine. Heat oil of choice in a soup pot and add onion. Sauté until onion is translucent. Stir often. Peel ginger and chop in blender with 2 Tbsp. water, until liquefied. Add this mixture to soup pot and cook for one minute longer. Stir well.

Peel, trim and cut parsnips into 1" pieces. Chop in food processor or blender. Add potato and process the machine until ingredients are well blended. Use small amounts of bouillon to keep blade moving. Add remaining bouillon and milk product of choice, a little at a time, continuing to chop until soup-like consistency is achieved.

Transfer soup to pot. Warm slowly, stirring often, to just below simmer point. Season to taste with tamari.

Garnish: Chopped cilantro

Parsnip Soup Simple

1 Tbsp. oil of choice (canola or olive oil, margarine,
    butter)
1 bay leaf
1/4 cup finely chopped onion
1 large parsnip
1 steamed potato
4 cups vegetarian white stock (see Soup Base
    in Introduction)

Heat oil in soup kettle. Crush bay leaf and add to oil.
Add onion and sauté until onion is transparent.

Peel, trim and cut parsnip into 1" pieces. Chop in food
processor or blender. Add steamed potato and continue
chopping. Slowly add vegetarian white stock, continuing
to work machine until a pureed consistency is achieved.

Add puree to soup kettle. Warm slowly. Stir frequently.
Do not allow soup to reach a simmer. Season with salt
and pepper.

Garnish: Chopped fresh parsley

Parsnip Soup Spicy Style

1 Tbsp. oil of choice (olive, canola, margarine, butter)
1/2 small onion
1/2 Tbsp. mustard seed
1/2 Tbsp. cumin seed
1/2 Tbsp. fennel seed
3 medium parsnips
1 medium steamed potato
4 cups vegetable cooking water or plain water
1 Tbsp. miso
Tamari to taste

Heat oil of choice in a soup pot. Chop onion fine and sauté until golden. Stir often. In a dry frying pan roast mustard seeds, cumin seeds and fennel seeds for three minutes over high flame. Grind the roasted seeds in a food mill and add to the sautéed onion. Cook for one minute longer. Stir well.

Peel trim and cut parsnips into 1'' chunks. Chop in a food processor or blender. Quarter the potato and add it to the machine, continuing to chop until the ingredients are assimilated. Use small amounts of vegetable cooking water or plain water to keep blade moving. Add remaining liquid, pureeing until a soup-like texture is achieved.

Transfer soup to pot. Dissolve miso in 1/2 cup of soup and add to the pot. Warm the soup slowly to just below simmer point. Stir well. Add tamari to taste.

Garnish: Chopped green scallions

Pumpkin Soup Seminole Style

1 cup milk
1/4 tsp. thyme
1/2 tsp. nutmeg
1 tsp. sugar
1/2 tsp. salt
2 cups (when chopped coarsely) raw onion
2 cups (when chopped coarsely) raw pumpkin
1 medium steamed potato
1/2 large green pepper
1 large tomato
1 green onion
2 sprigs parsley
4 cups chicken-flavored bouillon

Pour milk into soup pot and add seasonings. Warm over low flame and allow to stand while preparing other ingredients.

Peel pumpkin and remove seeds. You can dry them in a 350° oven and enjoy pumpkin seeds. Cut pumpkin into 1'' chunks. Chop coarsely in food processor or blender. Check to be sure you have two cups. Add tomato, green pepper, green onion and parsley; continue chopping. Use a little bouillon to keep blade moving. Add remainder of bouillon, and puree until a soup-like texture has been achieved.

Combine soup with milk and seasonings. Warm slowly over low heat, stirring frequently. Do not allow soup to simmer. Adjust seasoning with additional salt, if desired.

Garnish: Diced green pepper

Pumpkin Soup Zesty Style

1/2 large onion
1/2 Tbsp. oil of choice (olive, canola, butter, margarine)
1 large garlic clove
1 tsp. curry powder
1/4 tsp. ground coriander
2 cups (when chopped coarsely) raw pumpkin
1 medium steamed potato
4 cups chicken-flavored broth
1 cup milk product of choice (soy milk, evaporated
    skim milk, milk, cream)
1/2 tsp. salt

Mince onion and garlic. Heat oil of choice in soup pot
and sauté onion and garlic until tender. Add milk
product of choice and seasonings; keep warm.

Peel pumpkin, remove seeds (save seeds to dry; toast at
350° in oven). Cut pumpkin into 1″ chunks. Chop
coarsely in food processor or blender and check to be sure
you have two cups. Add steamed potato and continue
chopping. Add bouillon, a little at a time, working the
machine until mixture is pureed.

Add puree to ingredients in soup pot. Warm, stirring
frequently, but do not allow soup to simmer.

Garnish: Chopped chives and sour cream or soy sour
cream

Red Bell Pepper Fennel Soup

1 Tbsp. oil of choice (olive, canola, margarine, butter)
1 medium onion
4 cloves garlic
1 tsp. fennel seeds
1 cup milk product of choice (soy milk, evaporated skim
    milk, milk, cream)
2 large red bell peppers
1 small fennel bulb
1 small steamed potato
3 cups vegetable bouillon
Tamari to taste

Heat oil of choice in a kettle. Chop onion and garlic
fine. Roast fennel seeds in a dry frying pan over high heat
for several moments. Grind roasted fennel seeds in a food
mill. Add onion, garlic and ground fennel seeds to kettle.
Sauté until onion and garlic are soft. Stir often. Add milk
product of choice to kettle and warm it. allow this
mixture to cool.

Remove seeds from peppers (save seeds to sprinkle on
salads or to sprinkle on this soup as garnish). Cut peppers
into 1" pieces. Chop in food processor blender.

Cut fennel bulb into 1" pieces (remove fennel sprigs
and save for garnish). Quarter potato. Remove stems from
parsley and discard. Add fennel, potato and parsley
sprigs to work bowl, continuing to process until
ingredients are well assimilated. Use a little bouillon to
keep blade moving. Add remaining bouillon, pureeing
until a soup-like consistency is achieved. Transfer soup to
kettle. Warm slowly, stirring frequently, until just below
simmer point.

Garnish: Red bell pepper seeds or fennel sprigs

Red Bell Pepper Soup Cream Style

1 medium onion
3 cloves garlic
1 jalapeño pepper
1-1/2 Tbsp. oil of choice (olive, canola, margarine,
     butter)
2 large red bell peppers
1 medium steamed potato
3 cups chicken-flavored bouillon
1 cup milk product of choice (soy milk, evaporated
     skim milk, milk, cream)
1 Tbsp. lime juice
Salt and pepper to taste

Heat oil of choice in a soup kettle. Chop onion, garlic and jalapeño pepper fine. Sauté until soft. Stir often.

Remove seeds and core from red peppers. Chop pepper in food processor or blender. Quarter potato and add it to machine, continuing to chop until ingredients are assimilated. Use a little bouillon to keep blade moving. Add remaining bouillon and milk product of choice, a little at a time, working the machine until a soup-like consistency is achieved.

Transfer soup to the kettle. Warm slowly, stirring often, to just below simmer point. Add lime juice and season to taste with salt and pepper.

Garnish: Chopped parsley and, if desired, sour cream or soy sour cream

Red Bell Pepper Tomato Soup

1-1/2 Tbsp. oil of choice (olive, canola, margarine,
    butter)
1" piece fresh ginger
1 medium onion
4 cloves garlic
1 tsp. cumin
1 tsp. coriander
1 Tbsp. lemon juice
1 tsp. chili powder
4 medium red bell peppers
4 medium tomatoes
1 medium steamed potato
Salt and pepper to taste

Heat oil of choice in soup pot. Chop ginger, onion and
garlic fine. Sauté until soft. Add cumin, coriander, lemon
juice and chili powder. Cook a few minutes longer. Stir
constantly.

Cut peppers into 1" chunks, discard seeds and core.
Chop in food processor or blender. Quarter tomatoes and
potato; alternately add them to work bowl, continuing to
chop until ingredients are assimilated. Add vegetable
bouillon, a little at a time, pureeing until a soup-like
consistency is formed.

Transfer soup to the pot and warm slowly, stirring
frequently. Do not allow soup to simmer. Season to taste
with salt and pepper.

Garnish: Kernels of fresh corn

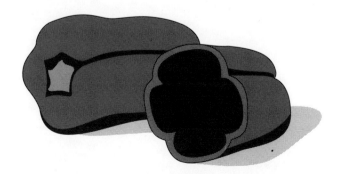

Spinach Soup Cream Style

1 Tbsp. oil of choice (olive, canola, butter, margarine)
1/2 small onion
1 bay leaf
1/8 tsp. ground cloves
1 cup milk product of choice (soy milk, evaporated
     skim milk, milk, cream)
2 bunches spinach
1 medium steamed potato
2 sprigs parsley
3 cups chicken-flavored bouillon
Salt and pepper to taste

Heat oil of choice in a soup pot. Chop onion fine and sauté until golden. Add milk product of choice, cloves and bay leaf. Scald milk. Allow to cool and remove bay leaf.

Clean spinach and discard stems. Chop spinach and parsley in a food processor or blender. Add potato and continue to work the machine. Add bouillon, a little at a time, and puree until a soup-like texture is achieved.

Transfer soup to pot and warm, stirring frequently. Do not allow the soup to come to a simmer. Season to taste with salt and pepper.

Garnish: Hard-boiled egg slices

Spinach Soup Curry Flavor

1 medium onion
1 Tbsp. oil of choice (olive, canola, margarine, butter)
1 tsp. curry powder
1/8 tsp. mace
1 lb. spinach
2 medium apples
1 medium steamed potato
4 cups chicken-flavored bouillon
Salt and pepper to taste

Chop onion fine. Heat oil of choice in a kettle. Add onion, curry powder and mace. Sauté until onion is soft. Stir frequently.

Discard thick stems of spinach. Chop spinach leaves in a food processor or blender. Quarter the apples and discard core and seeds. Add apples and boiled potato to machine and continue chopping until ingredients are assimilated. Use small amounts of bouillon to keep blade moving. Add remaining bouillon, and puree until a soup-like consistency is achieved.

Transfer soup to the kettle. Warm, stirring frequently, until just below simmer point. Season to taste with salt and pepper.

Garnish: Small thin pieces of apple

Spinach Soup Italian Style

1 Tbsp. olive oil
4 shallots
1 tsp. basil
2 Tbsp. tomato paste
5 sprigs parsley
5 Roma tomatoes
1 medium steamed potato
4 cups vegetable bouillon
Salt and pepper to taste

Chop shallots fine. Heat olive oil in a kettle. Add shallots, basil and tomato paste. Sauté until shallots are soft. Stir frequently.

Remove thick stems from spinach and discard. Chop spinach in food processor or blender. Add parsley, tomatoes and potato, continuing to puree until soup-like consistency is achieved.

Remove soup to the kettle. Warm, stirring frequently, until just below simmer point. Season to taste with salt and pepper.

Garnish: Cooked brown rice sprinkled with Parmesan cheese. Use soy Parmesan, if desired.

Spinach Soup Peanut Flavor

1 lb. spinach
1 medium Gravenstein apple
1/2 cup peanut butter
4 cups chicken-flavored bouillon
1 Tbsp. curry powder
Salt and pepper to taste

Wash spinach thoroughly. Remove tough ends of stems and discard. Chop spinach in a food processor or blender. Quarter apple, peel and remove seeds. Add apple to work bowl and continue chopping. Add peanut butter and work machine until ingredients are assimilated. Use small amounts of bouillon to keep blade moving. Add remaining bouillon and continue processing until a soup-like consistency is formed. Add curry powder and whirl machine once again.

Remove soup to kettle and warm slowly. Do not allow soup to simmer. Stir frequently.

Garnish: Millet, a yellow grain similar to rice with a much shorter cooking time (available in health food stores).

Bring 1 cup water to boil, add 1/2 cup millet and lower heat. Simmer millet in covered pot 15 minutes.

Spinach Soup Scandinavian Style

4 medium onions
1 Tbsp. oil of choice (olive, canola, margarine, butter)
1-1/2 lb. spinach
1 bunch fresh dill
1 medium steamed potato
4 cups chicken-flavored bouillon
Salt and pepper to taste

Chop onion finely and sauté, until golden, in oil of choice.

Remove stems from spinach and thoroughly clean leaves (for a thicker textured soup, use spinach leaves with stems intact). Chop in food processor or blender. Add sprigs of dill, but not stems. Quarter and blend potato with mixture, using small amounts of chicken-flavored bouillon. Add remaining bouillon, and puree until soup-like consistency is formed.

Transfer soup to pot and warm slowly, stirring frequently. Do not allow soup to come to simmer. Adjust seasoning to taste with salt and pepper.

Garnish: Grated lemon peel and sour cream or soy sour cream.

Spinach Soup Simple

1 Tbsp. margarine or butter
2 Tbsp. whole wheat flour
1 cup milk product of choice (soy milk, evaporated
    skim milk, milk, cream)
1/8 tsp. ground cloves
2 bunches spinach
4 scallions
3 cups chicken-flavored bouillon
Salt and pepper to taste

Heat margarine or butter in a soup pot. Add whole
wheat flour and stir until blended. Combine milk product
of choice with clove powder. Add to soup pot, a little at a
time, stirring continuously until well blended. Cook one
minute longer. Allow to cool.

Clean spinach and discard stems. Chop leaves in food
processor or blender. Add scallions and process. Add
bouillon, a little at a time, continuing to work machine
until a soup-like texture is formed.

Remove soup to pot. Warm, stirring frequently, but do
not allow soup to come to simmer. Season to taste with
salt and pepper.

Garnish: Cooked brown rice

Spinach Soup Pesto Flavor

5 shallots
4 teaspoons olive oil
Pinch of thyme
Pinch of oregano
1 medium steamed potato
1 Tbsp. pesto sauce (bottom of next page)
1 pound fresh spinach
1 medium zucchini
1/2 pound spinach noodles
Salt to taste

Chop shallots fine. Heat olive oil in a soup pot. Add chopped shallots, oregano and thyme. Sauté until shallots are golden brown.

Remove spinach stems and wash leaves. Puree spinach, a little at a time, in food processor blender. Add a little bouillon and continue pureeing. Cut zucchini into chunks and add a few chunks to spinach mixture; continue processing. Keep adding small amounts of bouillon to puree. When zucchini is absorbed in puree, add potato; continue working machine and adding bouillon until a soup-like consistency is achieved.

Add soup to sautéed shallots. Warm soup, stirring often, to serving temperature but do not allow it to simmer.

Cook noodles and mix with pesto. Add noodles to soup. Salt to taste.

Garnish: Grated Parmesan cheese or Soy Parmesan

Summer Squash Dill Soup

1 medium onion
1 Tbsp. oil of choice (olive, canola, margarine, butter)
5 medium summer squash
1 medium steamed potato
1 bunch dill
4 cups vegetable bouillon
Salt and pepper to taste

Chop onion fine. Heat oil of choice in a soup pot. Sauté onion until browned. Stir often.

Cut summer squash into 1″ chunks. Chop in food processor or blender. Discard stems from dill. Add leaves to work bowl and process. Add potato and continue to chop until ingredients are assimilated. Add a little vegetable bouillon to keep blade moving. Add remaining bouillon, continuing to work machine until a soup-like consistency is achieved.

Remove soup to pot. Warm, stirring often, to just below simmer point. Season to taste with salt and pepper.

Garnish: Thinly sliced celery

Pesto Recipe (See preceding page.)
3 cups fresh basil leaves, firmly packed
3 large cloves garlic
1/2 cup olive oil
3/4 cup Parmesan cheese or pine nuts

Put basil and cheese into work bowl and process until mixture is puree consistency. Add garlic and, with machine running, add oil a little at a time. Process until smooth.

Summer Squash Soup Colombian Style

1 small onion
2 cloves garlic
2 Anaheim peppers
1-1/2 Tbsp. olive oil
2 tsp. cumin
4 medium summer squash
10 sprigs cilantro
1 small carrot
1 small steamed potato
4 cups chicken-flavored bouillon
Salt and pepper to taste

Chop onion, garlic and peppers fine. Heat olive oil in soup pot and sauté until onion is translucent. Add cumin and stir well. Cook one minute longer.

Cut squash into 1" pieces. Chop in food processor or blender.

Discard stems from cilantro and add leaves to the work bowl, processing the machine. Cut carrot into 1" pieces. Add carrot and potato, continuing to chop until ingredients are assimilated. Use small amounts of bouillon to keep blade moving. Add remaining bouillon and process until a soup-like consistency is achieved.

Transfer soup to pot. Warm, stirring frequently, to just below simmer point. Season to taste with salt and pepper.

Garnish: Fresh corn kernels

Summer Squash Soup Cream Style

1 bunch scallions
1 Tbsp. margarine or butter
1 cup milk product of choice (milk, cream, soy milk,
    evaporated skim milk)
1/4 tsp. nutmeg
5 large summer squash
1 medium steamed potato
3 cups chicken-flavored bouillon
Salt and pepper to taste

Cut green ends of scallions off white tip and save for
garnish. Chop white tips fine. Heat margarine or butter in
soup pot and sauté scallion tips until they are tender.
Add milk product of choice and nutmeg; cook one minute
longer.

Cut summer squash into 1" chunks. Chop in food
processor or blender. Add potato, continuing to chop
until ingredients are assimilated. Use a little bouillon to
keep blade moving. Add remaining bouillon and process
until a soup-like consistency is achieved.

Remove soup to pot. Warm, stirring frequently, to just
below simmer. Season to taste with salt and pepper.

Garnish: Green ends of scallions, sliced thin

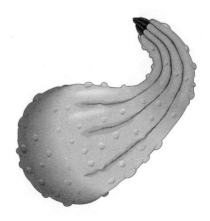

Summer Squash Soup Simple

1 large onion
1-1/2 Tbsp. oil of choice (olive, canola, margarine,
    butter)
6 medium summer squash
8 sprigs parsley
1 medium steamed potato
4 cups chicken-flavored bouillon
1 Tbsp. Spike (seasoning found in health food stores)

Chop onion fine. Heat oil in soup kettle and sauté onion
until soft. Stir often.

Cut squash into 1" pieces. Remove stems from parsley
and discard. Quarter potato. Chop squash in food
processor or blender. Add parsley and potato, continuing
to chop until ingredients are assimilated. Use small
amounts of bouillon to keep blade moving. Add
remaining bouillon and process until a soup-like
consistency is achieved.

Transfer soup to pot. Warm, stirring frequently, to just
below simmer point. Season to taste with Spike.

Garnish: Fresh dill

Summer Squash Soup Apple Flavor

1 Tbsp. oil of choice (olive, canola, margarine, butter)
1/2 small onion
2 cloves garlic
1/2" piece of ginger
2 cloves garlic
1/2 tsp. cumin
1/2 tsp. coriander
1/2 tsp. dry mustard
1/4 tsp. cinnamon
1 tsp. salt
6 medium summer squash
1 medium steamed potato
1 large apple
4 cups leftover water from steaming vegetables
    or plain water
1 Tbsp. lemon juice

Chop onion and garlic fine. In soup kettle with oil of choice, add cumin, coriander, mustard, cinnamon, salt and cayenne. Sauté onion and garlic until soft. Stir often.

Peel ginger. Place in blender with 1 Tbsp. water and blend. Add blended ginger to soup pot.

Cut summer squash into 1" pieces. Quarter potato and apple, discarding apple core and seeds. Chop squash in food processor or blender. Add potato and apple, continuing to chop until ingredients are assimilated. Use small amounts of liquid to keep blade moving. Add remaining liquid, a little at a time, working the machine until a soup-like consistency is achieved.

Transfer soup to pot. Warm, stirring frequently, to just below simmer point. Add lemon juice. Taste and adjust seasoning with additional salt and cayenne, if desired.

Garnish: Chopped cilantro

Tomato Carrot Soup

1-1/2 Tbsp. oil of choice (olive, canola, margarine,
   butter)
2 leeks
1 tsp. tarragon
2 Tbsp. tomato paste
5 medium, very ripe tomatoes
4 medium carrots
1 small steamed potato
4 cups vegetarian white stock (see Introduction)
Salt and pepper to taste

Heat oil of choice in kettle. Slice white part of leek and
sauté until tender. Add tarragon and tomato paste. Cook
one minute longer.

Quarter tomatoes and potato. Cut carrots into 1"
chunks. Chop tomatoes, potato and carrots in food
processor or blender until ingredients are combined. Use
a little vegetarian white stock to keep blade moving. Add
remaining stock, slowly pureeing until a soup-like
consistency is achieved.

Transfer soup to kettle. Warm, stirring frequently, to
just below simmer. Season to taste with salt and pepper.

Garnish: Cooked millet (See Spinach Soup Peanut Flavor
recipe for cooking directions.)

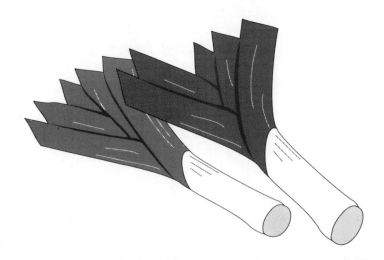

Tomato Peanut Butter Soup

1 medium onion
1-1/2 Tbsp. oil of choice (olive, canola, margarine,
    butter)
2 Tbsp. tomato paste
1 tsp. tarragon
1 tsp. cumin
2 celery stalks
6 medium very ripe tomatoes
3/4 cup peanut butter
4 cups leftover vegetable cooking water or plain water
Salt and pepper to taste

Chop onion fine. Heat oil of choice in soup pot. Sauté
onion until browned. Add tomato paste, tarragon and
cumin. Cook for one minute longer. Stir well.

Cut celery stalks into 1" pieces and chop in food
process or blender. Quarter tomatoes and add to work
bowl, continuing to process the machine. Add the peanut
butter and chop until ingredients are assimilated. Add the
vegetable cooking water, a little at a time, pureeing until a
soup-like consistency is achieved.

Transfer soup to pot. Warm, stirring often, just below
simmer point. Season to taste with salt and pepper.

Garnish: Raw peanuts

Tomato Pesto Soup

1 small onion
3 cloves garlic
1 Tbsp. olive oil
1/4 tsp. oregano
4 Tbsp. pesto*
6 very ripe medium tomatoes
1 large steamed potato
4 cups leftover vegetable cooking water or plain water
Salt and pepper to taste

*See Spinach Pesto Soup for pesto recipe

Chop onion and garlic fine. Heat olive oil in a soup pot and sauté onion and garlic until lightly browned. Add oregano and pesto. Stir well.

Quarter tomatoes and chop in food processor or blender. Add potato to work bowl, chopping ingredients until blended. Add vegetable cooking water, a little at a time, pureeing until a soup-like consistency is achieved.

Transfer soup to pot. Warm over low heat, stirring frequently; do not allow to simmer. Season to taste with salt and pepper or additional pesto.

Garnish: Cooked vermicelli and grated Parmesan cheese or soy Parmesan

Tomato Soup Cream Style

1 medium onion
1-1/2 Tbsp. oil of choice (olive, canola, margarine,
    butter)
1 tsp. tarragon
2 tbsp. tomato paste
6 medium very ripe tomatoes
1 medium potato steamed
4 cups chicken-flavored bouillon
4 Tbsp. tapioca
1/2 cup buttermilk or 1/2 cup soy milk
    with 1/4 Tbsp. lemon juice
1 tsp. sugar
Salt and pepper to taste

Heat oil of choice in kettle. Chop onion fine and sauté
until translucent. Add tarragon and tomato paste; cook
one minute longer. Add chicken-flavored bouillon and
tapioca; cook for half an hour. This creates a delicious
cream soup base. Allow to cool.

Quarter tomatoes and potato. Chop in food processor or
blender, alternately adding tomato and potato. Add
cream soup base, a little at a time, pureeing until a
soup-like consistency is achieved. Add buttermilk or soy
milk and lemon juice and sugar. Whirl machine once
again.

Remove soup to kettle. Warm over low heat, stirring
frequently. Do not let soup reach simmering point. Season
to taste with salt and pepper.

Garnish: Chopped onion chives

Tomato Soup Onion Flavor

1 medium onion
1 Tbsp. margarine or butter
1 tsp. marjoram
1/4 cup sherry
5 very ripe, medium tomatoes
1 stalk celery
5 fresh basil leaves
1 small steamed potato
4 cups onion-flavored stock (see Introduction
    or use powdered mix with water)
Salt and pepper to taste

Heat margarine or butter in kettle. Slice onion thin and sauté until translucent. Add marjoram and sherry. Cook one minute longer. Stir well.

Quarter tomatoes. Chop in food processor or blender. Cut celery into 1" pieces. Add celery, basil leaves and potato to work bowl, chopping until ingredients are combined. Add onion-flavored stock, a little at a time, pureeing until a soup-like consistency is achieved.

Transfer soup to kettle. Warm to serving temperature, stirring frequently, but do not allow to simmer. Season to taste with salt and pepper.

Garnish: Egg-onion matzo

Tomato Soup
Pineapple Flavor

5 medium, very ripe tomatoes
1 small steamed potato
2 cups pineapple chunks (canned or very ripe fresh)
6 basil leaves
6 parsley leaves
3 cups leftover vegetable cooking water
    or plain water
Salt and pepper to taste

Quarter tomatoes and potatoes. Chop them in food
processor blender. Add pineapple chunks, basil leaves
and parsley leaves, continuing to chop. Use a little
vegetable cooking water to keep blade moving. Add
remaining liquid, a little at a time, pureeing until a
soup-like consistency is achieved.

Transfer soup to a kettle. Warm, stirring frequently, to
just below simmer point. Season to taste with salt and
pepper.

Garnish: Chopped mint

Tomato Soup Saffron Flavor

1-1/2 Tbsp. oil of choice (olive, canola, margarine,
    butter)
1 medium onion
4 cloves garlic
1/2 tsp. thyme
1/2 tsp. sage
1/2 tsp. cumin
1/4 tsp. saffron
2 Tbsp. tomato paste
1/4 cup sweet vermouth
4 large tomatoes
1 medium potato steamed
4 cups chicken-flavored bouillon
1/2 tsp. sugar
Salt and pepper to taste

Chop onion and garlic fine. Add to oil of choice with
thyme, sage, cumin. Heat in a soup pot until onions and
garlic are slightly brown. Stir frequently. Rub saffron
between fingers and combine with 1 tsp. hot water. Add
to sauté along with tomato paste and vermouth. Cook for
three minutes longer. Allow to cool.

Quarter tomatoes and potato. Chop in a food processor
or blender until they are assimilated. Add the bouillon, a
little at a time, continuing to process the machine until a
soup-like consistency is achieved.

Transfer the soup to the soup pot. Warm, stirring
frequently, just below the simmering point. Add sugar,
salt and pepper to taste.

Garnish: Cooked rice

Tomato Soup Simple

1 Tbsp. oil of choice (olive, canola, margarine, butter)
1/2 medium onion
3 garlic cloves
1 bay leaf
1/2 tsp. paprika
Salt and pepper to taste
5 medium, very ripe tomatoes
1 medium steamed potato
5 sprigs parsley
4 cups vegetable bouillon
1-1/2 tsp. sugar

Heat oil of choice in a kettle. Chop onion and garlic fine. Sauté until slightly brown. Crush bay leaf in a food mill. Add crushed bay leaf and paprika to sauté and cook for two minutes longer. Stir well.

Quarter tomatoes and potato. Remove the stems from the parsley. Chop tomatoes, potato and parsley leaves in food processor or blender. Use a little vegetable bouillon to keep blade moving. Add remaining bouillon, pureeing until a soup-like consistency is achieved.

Transfer soup to kettle. Add sugar. Heat just below simmering point, stirring often. Season to taste with salt and pepper.

Garnish: Raw asparagus tips

Vegetable Soup Basque Style

1 medium onion
1-1/2 Tbsp. olive oil
1/4 tsp. thyme
1 small green pepper
1 stalk celery
1 small carrot
1 zucchini
6 fresh basil leaves
10 sprigs parsley
4 sprigs dill
1 small steamed potato
4 cups chicken-flavored bouillon
Salt and pepper to taste

Slice onion thin. Heat oil of choice in a soup kettle. Sauté onion until transparent. Add thyme and stir well.

Remove the seeds from green pepper and discard. Cut green pepper, celery, carrot and zucchini into 1″ chunks. Chop these ingredients a few at a time in a food processor or blender. Remove the stems from parsley and dill. Add parsley, dill, basil and potato to the work bowl, and process until the ingredients are assimilated. Use a little of bouillon to keep blade moving. Add remaining bouillon slowly, continuing to chop until a soup-like consistency is achieved.

Transfer soup to kettle. Heat, and bring soup to just below the simmer point. Stir often. Season to taste with salt and pepper.

Garnish: **Crusty Tomato Bread**

Soak 1 oz. dried tomatoes in 1/2 cup olive oil overnight. Spread this mixture on day-old slices of whole wheat sourdough bread and broil in oven for a few minutes. Cut into 2″ pieces.

Vegetable Soup Japanese Style

4 stalks celery
1/2 small onion
3 small carrots
1 medium potato
1/2 lb. asparagus
4 cups chicken-flavored bouillon
1/4 cup mirin (Japanese wine)
Tamari to taste

Cut carrots and celery into 1" pieces. Break off tough ends of asparagus (save tough ends for salads and stir-frying). Cut asparagus into 1" pieces. Chop celery, carrots and asparagus in food processor until fine. Add onion. If the machine gets stuck, add small amounts of the bouillon. Continue to work the machine, while adding the bouillon, a little at a time, until a soup-like consistency is formed.

Pour this mixture into a soup pot. Add the mirin and tamari. Heat, stirring frequently, but do not allow soup to come to simmer.

Garnish: **Toasted Sesame Seeds**

1 Tbsp. sesame seeds for each bowl of soup. Toast sesame seeds in a dry pan for about five minutes. Shake the pan often to keep seeds from burning.

Vegetable Soup
Mexican Style

1 Tbsp. oil of choice (olive, canola, margarine, butter)
1 medium onion
3 cloves garlic
2 medium carrots
2 stalks celery
1 medium green bell pepper
1 medium red bell pepper
1 medium tomato
8 sprigs cilantro
1 medium steamed potato
4 cups chicken-flavored bouillon
1/2 cup mild salsa

Heat oil of choice in soup kettle. Chop onion and garlic fine and sauté until they are lightly browned. Stir often. Add salsa.

Cut carrots, celery, green pepper and red pepper into 1" chunks. Quarter tomato and potato. Remove the stems from the cilantro and discard. Chop carrots, celery, green pepper and red pepper in food processor or blender. Use a little bouillon to keep blade moving. Add tomato, potato and cilantro leaves, continuing to chop until ingredients are assimilated. Add remaining bouillon, a little at a time, working the machine until soup-like consistency is achieved.

Transfer soup to kettle. Warm slowly to just below the simmer point, stirring often. Taste and adjust the seasoning with additional salsa, if desired.

Garnish: **Fried Tortilla Strips**

Tear tortillas into 1" strips. Coat the bottom of frying pan with little olive oil, and, over high heat, fry the tortilla strips quickly. Stir constantly so they will not burn. Add grated Cheddar cheese or soy Cheddar, if desired.

Vegetable Soup Swiss Style

1/2 medium onion
1 Tbsp. oil of choice (olive, canola, margarine, butter)
1/2 tsp. dry mustard
1/4 tsp. thyme
3 medium carrots
1 medium celery root
1 medium steamed potato
4 cups chicken-flavored bouillon
1 can beer
1 cup grated Swiss cheese
Salt and pepper taste

Chop onion fine. Heat oil of choice in a soup pot. Add onion, mustard and thyme and cook until onion is slightly brown. Stir frequently.

Cut carrots and celery root into 1″ chunks. Chop carrots and celery root in food processor or blender. Quarter steamed potato and add to mixture, continuing to chop. Add small amounts of bouillon to keep blade moving. Add remaining bouillon and beer, continuing to work machine, until a soup-like consistency is achieved.

Transfer soup to pot. Add grated cheese. Warm slowly, stirring constantly. Do not allow soup to reach a simmer. Season to taste with salt and pepper.

Garnish: Pretzels

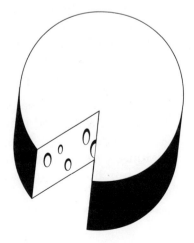

Vegetable Soup Portuguese Style

5 cups chicken-flavored bouillon
2 Tbsp. tomato paste
2 cloves garlic
1 cup elbow macaroni
1/2 lb. asparagus
1 scallion
2 small carrots
1/2 cup broccoli florets
1 small zucchini
6 sprigs parsley
1 small steamed potato
Salt and pepper to taste

Place 4 of the 5 cups of bouillon in a soup pot; reserve 1 cup to use later. Add tomato paste and stir until dissolved. Mince garlic and add to pot. Bring to a boil. Add macaroni and cook until soft. Allow to cool.

Break tough ends off of the asparagus (save for stir-frying). Chop spears in a food processor or blender. Cut carrots and zucchini into 1" pieces and add a few at a time to the machine, continuing to chop. Use a little of the fifth cup of bouillon to keep blade moving. Add potato, parsley and sliced scallion. Add remainder of bouillon and continue to work machine until a puree is formed.

Transfer puree to soup pot containing cooked macaroni. Warm, stirring frequently, but do not bring soup to a simmer. Season to taste with salt and pepper.

Garnish: **Sprouted Garbanzo Beans** (prepare in advance)

Soak garbanzo beans in water to cover overnight. Discard water and place beans in a colander to sprout (takes one or two days, depending on room temperature). Rinse several times a day. (Sprouted garbanzo beans can be purchased in many health food markets.)

Vegetable Soup Italian Style

1 Tbsp. olive oil
1 small onion
2 cloves garlic
1 tsp. basil
1/2 tsp. oregano
2 medium size carrots
1 large tomato
1/2 cup shelled peas
8 sprigs parsley
1 small steamed potato
4 cups vegetable bouillon
Salt and pepper to taste

Heat olive oil in soup pot. Chop onion and garlic fine; add to pot along with basil and oregano. Sauté until onions and garlic are lightly browned. Stir often.

Quarter tomato and potato; chop in food processor or blender until they are thoroughly assimilated.

Cut carrots into 1" chunks. Remove stems from parsley. (Do not use the stems in the soup—but they are great to munch on; refreshing and nutritious!)

Add carrot chunks and parsley, a few at a time, to work bowl, using a little bouillon to keep blade moving. Add remaining bouillon and continue pureeing until a soup-like consistency is achieved.

Transfer soup to pot and warm slowly, stirring often. Do not allow soup to come to a simmer. Season to taste with salt and pepper.

Garnish: Grated hard-boiled egg sprinkled with Parmesan cheese or soy Parmesan

Yam Soup Apple Flavor

4 shallots
1 Tbsp. oil of choice (olive, canola, margarine, butter)
1/4 tsp. nutmeg
1/4 tsp. cardamom
1/4 tsp. powdered ginger
1 cup apple juice
1 tsp. lemon juice
2 small yams
1 medium apple
3 cups vegetable bouillon
Salt and pepper

Mince shallot. Heat oil of choice in a soup pot. Sauté shallots until tender. Add nutmeg, cardamom, ginger and cook for one minute longer. Stir well. Add apple juice and lemon juice. Cook for a few minutes until heated through.

Peel yams and apple. Cut them into 1" chunks. Chop in a food processor or blender until they are combined. Use a little of the bouillon to keep blade moving. Add remaining bouillon slowly, continuing to process the machine until a soup-like consistency is achieved.

Remove soup to the pot. Warm, stirring often, and bring soup to just below simmer point. Season to taste with salt and pepper.

Garnish: Walnuts

Yam Almond Butter Soup

1 small onion
2 cloves garlic
1 Tbsp. oil of choice (olive, canola, margarine, butter)
1 tsp. cardamom powder
3 small yams
3/4 cup almond butter
4 cups vegetable bouillon
Salt and pepper to taste

Heat oil of choice in a soup kettle. Chop onion and garlic fine; sauté until they are soft. Add cardamom powder. Stir well.

Peel yams and cut into 1" chunks. Chop in a food processor or blender. Add almond butter and process. Add bouillon, a little at a time to keep blade moving. Add remaining bouillon slowly, continuing to chop until a soup-like texture is achieved.

Transfer soup to kettle. Warm to serving temperature, stirring frequently. Do not allow soup to simmer. Season to taste with salt and pepper.

Garnish: Slivered toasted almonds

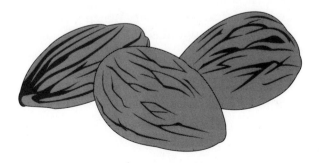

Yam Soup Indonesian Style

1-1/2 Tbsp. olive oil
1 medium onion
4 cloves garlic
1 tsp. curry powder
1/2 tsp. coriander powder
1/2 tsp. cardamom seeds
1/2 tsp. fennel seeds
1/2 tsp. cumin seeds
3 medium yams
1 cup spinach leaves
4 cups chicken-flavored bouillon
1/2 cup yogurt or soy yogurt
Tamari to taste

Heat olive oil in a kettle. Chop onion and garlic fine; sauté until lightly brown. Add curry powder and coriander powder. Stir well.

In a dry frying pan over high heat, roast cardamom seeds, fennel seeds and cumin seeds. Grind roasted seeds in a food mill and add to the sauté. Cook one minute longer. Stir thoroughly.

Peel yams and cut into 1" pieces. Chop in a food processor or blender. Remove the tough stems from spinach. Add to the work bowl and process. Add chicken-flavored bouillon slowly, continuing to chop until a soup-like consistency is achieved. Add yogurt and whirl the machine once again.

Transfer soup to kettle. Warm slowly over low heat. Do not allow soup to reach a simmer. Season to taste with tamari.

Garnish: Thinly sliced raw mushrooms and apricot chutney

Yam Soup Simple

1/2 small onion
3 cloves garlic
1 Tbsp. dried dill
1 Tbsp. oil of choice (olive, canola, margarine, butter)
3 large yams
3 medium carrots
4 cups chicken-flavored bouillon
Salt and pepper to taste

Chop onion and garlic fine. Heat oil of choice in a soup pot. Add onion, garlic and dill. Sauté until the onion and garlic are soft. Stir well.

Peel sweet potatoes and cut into 1" chunks. Cut carrots into 1" chunks. Chop sweet potatoes and carrots in a food processor or blender. Use a little of the bouillon to keep the blade moving. Add the remaining bouillon, a little at a time, working the machine until a soup-like consistency is achieved.

Transfer soup to pot. Warm slowly, stirring frequently, to just below the simmer point. Season to taste with salt and pepper.

Garnish: Fresh dill

Yam Soup Cream Style

1/4 medium onion
2 cloves garlic
1 Tbsp. oil of choice (olive, canola, margarine, butter)
1-1/2 tsp. coriander
1/2 tsp. thyme
2 medium yams
2 stalks celery
1 small carrot
3 cups vegetable bouillon
1 cup buttermilk (or soy milk
    and 3/4 Tbsp. lemon juice)

Chop onion and garlic fine. Heat oil of choice in a soup pot. Add onion, garlic, coriander and thyme. Sauté until the onion and garlic are soft. Stir well.

Peel sweet potatoes and cut into 1" chunks. Cut celery and carrot into 1" chunks. Chop yams, celery and carrot in a food processor or blender. Use a little of the bouillon to keep the blade moving. Add the remaining bouillon and buttermilk (or soy milk combined with lemon juice), a little at a time, continuing to process until a soup-like consistency is achieved.

Remove soup to pot and warm slowly to just below the simmer point. Stir often. Season to taste with salt and pepper.

Garnish: Chopped cilantro

Zucchini Chowder

1 small onion
1 Tbsp. oil of choice (olive, canola, margarine, butter)
1 tsp. oregano
1 cup milk product of choice (cream, milk, soy milk,
    evaporated skim milk)
3 medium zucchini
1 small red bell pepper
1 small steamed potato
4 sprigs parsley
3 cups chicken-flavored bouillon
Salt and pepper to taste

Chop onion fine. Heat oil of choice in a kettle and sauté
onion until translucent. Add oregano and cook one
minute longer. Add milk product of choice and cook until
warm. Stir well.

Cut zucchini into 1" pieces. Remove seeds from red
pepper and discard. Add pepper, potato and parsley
leaves (discard or chew on stems) to work bowl,
continuing to process until ingredients are assimilated.
Use a little bouillon to keep blade moving. Add
remaining bouillon, chopping until a soup-like consistency is achieved.

Transfer soup to kettle. Warm to serving temperature,
stirring frequently. Do not allow soup to simmer. Season
to taste with salt and pepper.

Garnish: Raw corn kernels

Zucchini Soup Simple

1 Tbsp. oil of choice (olive, canola, margarine or butter)
4 cups tomato juice
1/8 tsp. marjoram
1/8 tsp. thyme
4 small zucchini
1 medium steamed potato
Salt and pepper to taste

Heat oil in soup kettle. Add chopped onion, basil, thyme, marjoram; sauté until onion is slightly browned.

Quarter zucchini and puree in food processor or blender. Add steamed potato and continue pureeing. Add tomato juice, a little at a time; continue working machine until a soup-like texture is achieved.

Warm slowly, stirring frequently. Do not let soup reach a simmer. Season with salt and pepper to taste.

Garnish: Popcorn

Zucchini Soup Italian Style

1/2 onion
3 cloves garlic
1 Tbsp. dried basil
1/2 tsp. dried oregano
1 Tbsp. oil of choice (olive, canola, margarine or butter)
3 medium zucchini
1 medium carrot
1 medium tomato
1 medium steamed potato
4 cups tomato base bouillon (see Introduction)
Salt and pepper to taste

Mince onions and garlic. Heat oil of choice in a soup kettle. Add onions, garlic, basil and oregano. Sauté until onions and garlic are lightly browned. Stir frequently.

Cut zucchini and carrot into 1" chunks. Quarter the tomato and potato. Chop tomato in a food processor or blender. Add zucchini, potato and then the carrot. Continue chopping, using small amounts of bouillon to keep blade moving. Add remaining bouillon, and puree until a soup-like consistency is formed.

Remove soup to the kettle. Warm slowly, stirring frequently. Do not let the soup reach a simmer. Season to taste with salt and pepper.

Garnish: Cooked elbow macaroni and grated Parmesan cheese or soy Parmesan, if desired.

Zucchini Soup Mediterranean Style

1 cup garbanzo beans
5 cups water
1 bay leaf
1 medium onion
4 cloves
2 tsp. marjoram
2 Tbsp. cumin
1/8 tsp. cayenne
6 medium zucchini
3 Tbsp. lemon juice
Salt and pepper to taste

Rinse beans and discard any foreign objects. Soak beans overnight in just enough water to cover. The next day, discard soaking water. Place beans in a kettle with 5 cups of fresh water and a bay leaf. Bring to a boil, cover and simmer two hours or until tender. Discard bay leaf. Chop cooked beans in a food processor or blender, adding cooking liquid a little at a time. This will be the soup base.

Chop onion and garlic fine. Heat olive oil in a soup kettle. Add onion, garlic and spices. Sauté until onion and garlic are soft. Stir constantly.

Cut ends off zucchini and discard. Cut zucchini into 1″ pieces and chop them in a food processor or blender. Add small amounts of the bean soup base to keep blade moving. Add remaining soup base and continue chopping until zucchini is pureed and soup has a very smooth texture.

Combine soup with sautéed onions and garlic. Heat just below the simmer point, stirring frequently. Add lemon juice. Season to taste with salt and pepper.

Garnish: Sliced, pitted olives and, if desired, crumbled goat cheese

Zucchini Soup Pistou Flavor

4 large cloves garlic
4 Tbsp. tomato paste
10 fresh basil leaves
1/2 cup Parmesan cheese or soy Parmesan
1/2 cup olive oil
6 sprigs parsley
3 medium zucchini
1 small steamed potato
4 cups leftover vegetable cooking water, or plain water
Salt and pepper to taste

Press garlic in a garlic press. Chop basil leaves and parsley leaves fine. Add garlic, chopped basil, parsley, Parmesan cheese and tomato paste to olive oil. Blend well.

Cut zucchini and potato into 1" chunks. Chop in a food processor or blender until assimilated. Use a little of the leftover vegetable cooking water to keep blade moving. Add remaining vegetable cooking water and continue to work machine until a soup-like consistency is achieved.

Transfer soup to a kettle. Warm slowly over low heat, stirring frequently. Add already prepared pistou flavoring. Do not allow soup to reach a simmer. Season to taste with salt and pepper.

Garnish: Cooked white beans

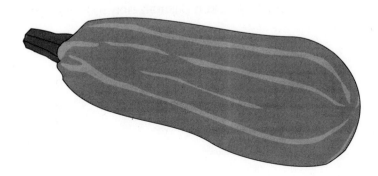

Zucchini Soup Spicy Style

1-1/2 tsp. canola or olive oil
1 bunch green onions
3 cloves garlic
1/4 tsp. powdered turmeric
1/4 tsp. powdered coriander
Pinch of cayenne
1/2 Tbsp. lemon juice
1/2 tsp. mustard seeds
1/2 tsp. cumin seeds
1 medium steamed potato
4 cups chicken-flavored bouillon
Salt to taste

Heat oil in a soup pot. Chop green onions and garlic fine. Sauté until tender. Add turmeric, coriander, cayenne and lemon juice. Stir well.

In a dry frying pan, roast mustard seeds and cumin seeds for a few minutes. Grind in a food mill and add to the sauté. Cook one minute longer. Stir thoroughly.

Trim zucchini and cut into 1" pieces. Chop in a food processor or blender. Add potato and process until ingredients are merged. Use a little of the bouillon to keep blade moving. Add remaining bouillon slowly, continuing to chop until a soup-like consistency is achieved.

Transfer soup to pot. Warm, stirring frequently, just below simmering point. Adjust seasoning with salt and additional cayenne.

Garnish: Cooked Basmati rice